The Soft Power of Construction Contracting Organisations

As is true across the industry, the non-technical skills possessed by construction organisations are key to their overall performance. In this study, the particular importance of optimising the so-called 'soft power' of organisations is addressed. Things like organisational culture, corporate learning behaviour, and building trust-based relationships with other stakeholders are seen as facets of a broader organisational capability, and the advantages of this strength are also explored.

The internationally conducted research behind this book identifies the importance of soft power to construction contracting organisations, and also shows what actions an organisation can take to improve its soft power. Readers of this book will gain new insights into effective management, from both inter- and intra-organisational perspectives. This unique and important book is essential reading for researchers and advanced students of construction management.

Sai On Cheung is a chartered quantity surveyor by profession. Before joining academia, he had substantial experience in contract administration with both consultant office and construction contracting organisations. Building on these experiences, Professor Cheung established the Construction Dispute Resolution Research Unit (CDRRU) and has developed research programmes in organisation issues in construction, contract and dispute management. He has also published widely in these areas. He has received two CIOB awards for his research in construction partnering and use of information technology to minimise dispute. A collection of his research in dispute management was published as the research monograph *Construction Dispute Research* in 2014. In addition, Professor Cheung is a recipient of the 2010 City University of Hong Kong Teaching Excellence Award.

Peter Shek Pui Wong is a Senior Lecturer at the RMIT University, overseeing the quantity surveying-related courses the university offers in Australia, Singapore and Hong Kong. Before embarking on his academic career, Dr Wong worked as a quantity surveyor with Rider Levett Bucknall Ltd, and was involved in a number of prestigious construction projects. He has a proven track record of publication in reputable construction management journals. He is a founding member of the Construction Dispute Resolution Research Unit (CDRRU) and the cluster leader of the greater China region of the Research Centre for Integrated Project Solutions of the RMIT University. He fosters a strong linkage between applied research and the industry.

Tak Wing Yiu is a Senior Lecturer in the Department of Civil and Environmental Engineering at the University of Auckland, and previously he worked in the field of quantity surveying. He is a founding member of the Construction Dispute Resolution Research Unit (CDRRU) and has conducted construction contracting, negotiation and mediation research for more than 10 years. He has built a proven track record, and most of his research outputs have been published in the top journals of these areas. Dr Yiu was a recipient of the 2012 ASCE JIADR Best Forum or Synopsis Paper Award and recognized as 2013 ASCE JIADR Outstanding Reviewer.

Spon Research

publishes a stream of advanced books for built environment researchers and professionals from one of the world's leading publishers. The ISSN for the Spon Research programme is ISSN 1940-7653 and the ISSN for the Spon Research E-book programme is ISSN 1940-8005

Published:

Free-Standing Tension Structures: From tensegrity systems to cable-strut systems
978-0-415-33595-9
W.B. Bing

Performance-Based Optimization of Structures: Theory and applications
978-0-415-33594-2
Q.Q. Liang

Microstructure of Smectite Clays and Engineering Performance
978-0-415-36863-6
R. Pusch and R. Yong

Procurement in the Construction Industry: The impact and cost of alternative market and supply processes
978-0-415-39560-1
W. Hughes *et al.*

Communication in Construction Teams
978-0-415-36619-9
S. Emmitt and C. Gorse

Concurrent Engineering in Construction Projects
978-0-415-39488-8
C. Anumba, J. Kamara and A.-F. Cutting-Decelle

People and Culture in Construction
978-0-415-34870-6
A. Dainty, S. Green and B. Bagilhole

Very Large Floating Structures
978-0-415-41953-6
C.M. Wang, E. Watanabe and T. Utsunomiya

Tropical Urban Heat Islands: Climate, Buildings and Greenery
978-0-415-41104-2
N.H. Wong and C. Yu

Innovation in Small Construction Firms
978-0-415-39390-4
P. Barrett, M. Sexton and A. Lee

Construction Supply Chain Economics
978-0-415-40971-1
K. London

The Soft Power of Construction Contracting Organisations

Edited by
Sai On Cheung, Peter Shek Pui Wong
and Tak Wing Yiu

Routledge
Taylor & Francis Group

LONDON AND NEW YORK

First published 2015 by Routledge

2 Park Square, Milton Park, Abingdon, Oxfordshire OX14 4RN
52 Vanderbilt Avenue, New York, NY 10017

Routledge is an imprint of the Taylor & Francis Group, an informa business

First issued in paperback 2019

British Library Cataloguing-in-Publication Data
A catalogue record for this book is available from the British Library

Library of Congress Cataloging in Publication Data
The soft power of construction contracting organisations / edited by Sai On
Cheung, Peter Shek Pui Wong and Tak Wing Yiu.
pages cm
Includes bibliographical references and index.
1. Construction industry—Management. 2. Construction contracts. 3.
Negotiation in business. I. Cheung, Sai On.
HD9715.A2S555 2015
692'.8068—dc23
2014033646

ISBN: 978-1-138-80528-6 (hbk)
ISBN: 978-0-367-37783-0 (pbk)

Typeset in Sabon
by FiSH Books Ltd, Enfield

To members and supporters of the Construction Dispute
Resolution Research Unit, with heartfelt thanks

Contents

x *Contents*

Figures

Tables

xiv *Tables*

Contributors

Chung Wai Keung is a chartered quantity surveyor and worked in an international consultant firm before City University of Hong Kong. His main research interests are in construction project networking and BIM management. His research aims to improve construction efficiency and cooperation through BIM deployment and promote collaborative project network to enhance BIM management.

Pui Ting Chow is a member of the Construction Dispute Resolution Research Unit. She received her PhD for a study in withdrawal during dispute negotiation from the City University of Hong Kong. The study won the Hong Kong Institute of Surveyors Best Dissertation Award (PhD category) in 2011. She has published in topics of negotiation strategies, trust development and dispute behaviour modelling. Her current research focuses on land use, innovation and sustainable development.

Wei Kei Wong is a chartered quantity surveyor. In 2004, she joined the Construction Dispute Resolution Research Unit of the City University of Hong Kong and completed a research study on trust for a Master of Philosophy. The study won the Hong Kong Institute of Surveyors Best Dissertation Award (MPhil category) in 2006. Miss Wong is now a contracts engineer in a railway operator in Hong Kong.

Wai Yin Wu completed her BSc (Hons) degree in Quantity Surveying at the City University of Hong Kong. In early 2007, she completed her Master of Philosophy entitled "Organization culture in construction industry". Her research has been shared with the construction community through publications in leading construction project management journals. She is now a quantity surveying officer in a department of the Hong Kong Government.

Ka Ying Chan had solid experience in quantity surveying practice before joining the Construction Dispute Resolution Research Unit of the City University of Hong Kong. Her principal research is on ways to build trust in construction contracting. Her current research studies include organizational capability for innovations. She is a co-editor of a catalogue of Hong Kong construction innovations published the Unit.

Hoi Yan Pang received her Master of Philosophy in 2010. The study examines the occurrence likelihood of construction disputes and won the Hong Kong Institute of Surveyors Best Dissertation Award (MPhil category) in 2011. In addition, she has worked on research in trust inventory, organisational culture and construction dispute. She organized an urban design competition and published a handbook for the topic.

Foreword

Construction is a huge business around the world with approximately one in ten people earning their living by working across the value chain in the delivery of new projects, ranging from small housing extensions to megaprojects, such as the upgrading of the Panama Canal or the building of the Kingdom Tower in Saudi Arabia over one kilometre high. Whilst the big projects grab the headlines, there is a huge amount of work in maintaining the existing built environment facilities. Construction is changing very fast, construction companies must react quickly to change or they face extinction. Clients want their projects cheaper with better value for money, they also want faster delivery, more certainty, higher quality, improved attention to safety and health, and to minimise their risk exposure. Technology has changed the materials used in the industry, and information and communication technology has changed the way projects are designed. Building information modelling is helping to improve the way information is used and will be transformational. Regulations must continually adapt to new realities of governance and standards. But, it is the people and the soft skills that make the companies and the industry work. Pieces of paper and regulations do not build projects or companies, they are necessary, it is the people that ensure projects can be delivered to meet the client expectations.

Because of the complex and diverse nature of construction, things do go wrong and disputes occur. Settling disputes takes a long time and is costly in financial, emotional, relationship, and business terms. Understanding aggressive behaviour and conflict means that a better understanding of people and organisations is needed.

Against this backdrop, this book on soft power meets a valuable need, it stimulates new ideas and new approaches to the management of a construction business, it suggests ways to improve and be more competitive. Most importantly, it encourages us all to think differently about what we do and how we do it. It is well written by a variety of authors who show they have really thought about the topics of performance, productivity, relationships, disputes, and trust in the industry. People are the backbone of the industry, there is a realisation of the importance of the soft factors that shape an organisation and influence behaviour, values, and ultimately, performance.

Soft power is an interesting term that reflects the importance of putting people and the soft issues at the heart of a business.

The keywords that describe the book are competitiveness, collaboration, culture, co-operation, understanding, creativity, learning, and trust. The authors are well known for their work in the area of trust and relationships. Competitiveness is very complex and multi-faceted. Twenty years ago being the lowest bidder on a project secured the work; in the changing construction industry of today that is not enough. Companies must prove they are ethically responsible, care about the environment, have safe working practices, and have employees who behave professionally and deliver on their promises. The best companies in the world are acutely aware of the need to become more professional in project delivery and relationship building. Organisational culture plays an important role, as does driving out inefficiencies and minimising disputes.

The book is a good balance between theory and practice. It uses research results to back up some of the ideas being put forward in the chapters. It will help the practitioner and the student to think again about how soft skills and power can be better understood. It is multi-disciplinary and uses economic theory with psychology, sociology, combined with the traditional construction disciplines.

The chapter on analysing aggressive cooperation drivers in construction organisations is fascinating. All companies think of themselves as fair, reasonable, harmonious, and collaborative, yet the chapter makes the reader think again. Contracts are fine when everything is working as planned, yet when something goes wrong, companies reach for the contract to settle any disagreements and aggression can manifest itself. With more international joint ventures, this could be a very important factor in the decision to joint venture with a company from overseas. The chapters on organisational learning provide an interesting insight into how companies really do learn. Companies need to recognise that theory can play an important role in shaping their ideas; catastrophe theory and complexity theory may not be in their everyday vocabulary, but it is when projects do not go according to plan.

I commend the book and the authors, I hope that this is the first step in a journey to understand how soft power is such an important factor and influence in the construction industry around the world.

Roger Flanagan
University of Reading, UK
Past President of the Chartered Institute of Building

Preface

The construction industry is well known to be very competitive. In recent years, construction costs have risen significantly, particularly for complex projects and environmentally friendly designs. To be competitive, many construction contracting organisations (CCOs) adopt a short-term approach of cost-cutting through outsourcing. However, this approach gives little incentive to enhance the long-term competitiveness of their organisations. This hampers the ability of construction contracting organisations to cope with new challenges. Other than technical competence, non-technical skills possessed by an organisation play an important role in project delivery. Thus it is important for CCOs to optimize the use of their resources. This book posits to identify the non-technology-based competitive edges of CCOs. These areas are called 'soft power', signifying the organisational strength of CCOs that differentiate them from other competitors. Soft power is not costly and would be treasured by project clients who value the ability of CCOs in delivering projects amicably.

This book is a collection of studies focusing on the ways construction contracting organisations can enhance competitiveness by developing organisational strength that distinguishes them from their competitors. Three types of ability are discussed: (i) ability to work collaboratively in a competitive environment; (ii) ability to be a learning organisation; and (iii) ability to minimise non-productive use of scare resources. These abilities are collectively termed as 'soft power'. Each chapter can also be read as a stand-alone research study. This book has four sections.

In the first section, competitiveness is examined from an organisational capacity-building perspective. The construct of soft power is discussed. CCOs with exceptional soft power will have a competitive edge over their competitors. Three types of ability are discussed: (i) ability to work collaborative in a competitive environment; (ii) ability to be a learning organisation; and (iii) ability to minimize non-productive use of scarce resources. Each type of ability is examined seriatim in three sections.

In the second section, whether collaborative working is feasible in competitive contracting environment is explored. Project performance is inevitably negatively affected by protracted disputes. The resulting deterioration of the relationship between contracting parties hampers cooperation; these negatively affect the organisation's overall performance. A trusting environment would help to bridge gaps, establish faith and synergize the strengths of members of an organisation. A conceptual framework of trust for construction contracting is offered in this book. The validity of the framework is again empirically tested. The element of trust is then extended to the situation of project dispute negotiation and partnering. The findings suggest that trust is instrumental in making significant contributions in enhancing the competitiveness of the construction organisations. The drivers for cooperation are also discussed.

The last chapter of this section discusses how a conducive, progressive and enduring culture underpins an efficient organisation. Organisational culture has several important functions. First, it conveys a sense of identity for organisation members. Second, it facilitates the generation of commitment. Third, it engenders stability. Fourth, it serves as a sense-making device that can guide and shape behaviour. This book offers an organisational culture framework to capture the cultural characteristics of contracting organisations. With this framework, how organisational culture influences construction project performance is hypothesised and empirically tested.

Organisational learning is the focus of the third section of the book. Organisational learning refers to a process of applying the imbibed knowledge for performance improvement. It is essential for construction-contracting organisations to sustain continuous improvement in the turbulent construction market. Unfortunately, construction-contracting organisations have been criticised of being incapable of learning. In this regard, this book firstly conceptualises how learning is practised within and among construction-contracting organisations with due attention to the very nature of the construction projects. The effect of organisational learning on performance is investigated. For this purpose, the theoretical anchors are examined and empirically tested. This offers insight about the common type of learning adopted by construction organisations and also choosing the appropriate type of learning to facilitate performance improvements. Furthermore, whilst it is apparent that construction-contracting organisations should strive to learn more effectively, their learning behaviours in projects have rarely been delved into. In many reported studies, performance-monitoring systems were purposefully developed for fostering contractors' organisational learning. Contractors were postulated as a rational entity that responds optimally to the feedback generated by the performance monitoring systems. Scholars have long argued such postulation is reflective to the true learning behaviour of the construction contractors. However, few previous studies have provided empirical support to demonstrate the construction contractors' learning responses to the performance feedback. In this book, the contractors' dichotomous responses to the project performance feedback are demonstrated by using a catastrophe model.

The fourth section of this book gets to the heart of a classic management issue: managing human influence in construction contracting. All projects are delivered by the people working on them. Managing people has always been a challenge in multi-disciplinary and highly complex projects. In fact, this challenge is often behavioural rather than technical in nature. Construction professionals need to have soft skills such as team-building, influencing, communication, etc. to achieve the project goals. However, these are seldom learnt as part of formal education process for construction professionals but rather through hard experience.

The dynamic changes of contracting behaviour under conflict situation are often the result of interactions among project team members. Chapter 8 discusses how contracting parties act differently and appropriately in different tension and conflict situations. The final two chapters introduce how Adam's Equity Theory can be applied to investigate the behavioural responses of contracting parties under inequitable situations. These situation happen when construction professionals are having face-to-face interactions, exchanges of information, making concessions or compromising. This study helps to understand the ways construction professionals respond to equitable or inequitable situations.

This book is written for academics, professionals and students working in construction project management. It will be an asset to those who wish to develop resilient construction contracting organisations. In addition to the practical outlook, the studies included in this book are firmly anchored on theoretical bases in the areas of psychology, economics and sociology. These theoretical approaches enrich the explanatory power and broaden the width of discussions. Multi-dimensional discussion on the subject stimulates readers' thinking. Furthermore, a good range of quantitative and qualitative research methods have been used for the studies. Strong theoretical bases coupled with robust research methodologies and input from professionals ensure the academic and practical value of the studies.

The authors are in debt to Professor Roger Flanagan for his inspiring Foreword. This book could not have been completed without the contributions of the members of the Construction Dispute Resolution Research Unit. The authors are also grateful to Miss Tina Chan for the production support.

Sai On Cheung
Peter Shek Pui Wong
Tak Wing Yiu
September 2014

Section A

Soft power as a form of competitive advantage

1 Soft power in construction contracting

Sai On Cheung, Ka Ying Chan and Pui Ting Chow

Introduction

The construction industry is one of the main contributors to the economy of many countries. For example, in Hong Kong the total gross value of construction works performed by main contractors in the first quarter (Q1) of 2014 has increased on the previous year by 12.7 per cent to HK$49.2 billion. The gross value of construction works performed by construction establishments covers the output of architectural and civil engineering works at construction sites, renovation, repair and maintenance of immobile structures and specialised construction-related job areas such as concreting, scaffolding, carpentry, plumbing and gas works (Census and Statistics Department 2014). Major infrastructural developments were planned in the millennium with the aim of raising the competitiveness of Hong Kong and to meet with the economic challenges. Infrastructure and housing are in great demand in developing economies; these markets offer exceptional opportunities for construction contracting organisations (CCOs). With the increase in project complexity and expectation from the stakeholders, construction contracting is now far more than simple coordination of trades. CCOs should be well equipped with adaptive technical and management capabilities in order to succeed in competitive construction contracting.

Unfortunately, the construction industry is often seen to be inefficient and confrontational, with endless arguments between the participating organisations; typically in a development project, other than the project client, a team of professionals working together to prepare the design. CCOs are primarily contractors who are charged with the responsibility of materialising the design. The construction contracting business environment is changing rapidly with stakeholder's raising expectations, rising concern over environmental issues and escalating material and labour costs. In addition, competition for construction works is fierce, especially in markets where competition is global. CCOs offering no competitive advantage will inevitably be unsuccessful. In what ways can CCOs develop competitive advantage? Ownership of advanced technology has been the primary means to achieve relative competitive advantage. Notable examples can be found in the heavy engineering and automobile industries. Nonetheless, for many small- and medium-sized CCOs, developing advanced technology may not be within their capacity. It would make better sense for them to look into their management abilities and establish distinctive organisational strength. In fact, the ability to achieve amicable completion has long been treasured by project clients. The same expectation extends to all CCOs. In this book, the soft power of CCOs are those non-technology-based edges. Competitive advantages are the natural result of CCOs with distinctive soft power in managing complex and conflict-laden construction contracting. Soft Power is mainly organisational and manifested in three types of ability: (i) to work collaboratively in a competitive environment; (ii) to continue to improve as a learning organisation; and (iii) to minimise non-productive use of scarce resources. Detailed discussion on each of the three types of ability is provided in Sections B, C and D respectively of the book.

Competition in construction contracting

Competition in construction contracting is fierce. Excessive competition will erode the profit of CCOs (Dess *et al.* 2012; Porter 1998). Five competitive forces have been identified by Porter (1998):

1   threat of new entrants;
2   bargaining power of buyers;
3   bargaining power of suppliers;
4   threat of substitute products and services;
5   rivalry among existing firms.

These forces determine the extent of competition and thus profitability.

Establishing barriers to new entrants is one commonly used means to reduce competition. However, instigating barriers is often viewed as anti-competitive and is prohibited under anti-trust policies. Construction contracting remains one of the most open and competitive businesses. Furthermore, the bargaining power of buyers is conventionally strong in most economies unless suppliers can develop unique irreplaceable capabilities. To excel among rival firms, CCOs should develop abilities that distinguish them from their competitors. For example, Fiol and Lyles (1985) suggested that organisations have to align themselves with the environment to remain competitive and innovative so that long-term survival and growth can be achieved.

In terms of creating competitive advantages, it is not uncommon for CCOs to employ a cost-cutting approach to improve their competitiveness. Cut-throat pricing has been the major means for CCOs to win projects in competitive bidding. Moreover, faced with the global trend of increasing project sophistication, concerns over environmental issues and escalation of material and labour cost, relying on cost-cutting alone leaves little room for long-term survival. Porter (1998) also identified this detrimental effect of cost competition. Competition in cost is highly unstable and quite quickly the entire industry becomes worse off as far as profitability is concerned. Cost-cutting not accompanied with efficiency gain does not generate genuine competitive edge.

From a broader perspective, it has been advocated that employers, especially the public sector, are the appropriate stakeholders to instigate operational efficiency through quality-focused procurement and contractual arrangements (CIRC 2001). The first step is to change the project award criteria. Although price remains the overriding consideration in tender award, the rising importance of non-price factors can also be evidently noted. For example, the government of Hong Kong Special Administrative Region (HKSARG) has revised the criteria of tender evaluation to reflect the changing expectation of project clients. In 1998, the weighting of price and technical score was 80 and 20 respectively (HKSARG 1998). As from 2002, the weighting of price and technical score was changed to 60 and 40 respectively (HKSARG 2002). The shift in the ratio highlights the growing concern on the ability of the contractor in amicably completing the project. Contractor's resources such as managerial staff and past performance are the scoring attributes for the technical score. In 2004, claim attitude was further included as a technical attribute in tender assessment (HKSARG 2004). The Hong Kong Housing Authority (HKHA 2011) includes ability to innovate as an award criterion. The rising importance of non-price factors that demonstrate the ability of the contractor in amicable project completion aptly supports the need of CCOs to develop soft power.

Soft power: a form of competitive advantage

Porter (1998: 5) suggested three generic competitive strategies to address the five competitive forces: (1) overall cost leadership; (2) differentiation; and (3) focus. Overall cost leadership refers to 'vigorous pursuit of cost reductions from experience, tight cost and overhead control' (35). However, while achieving cost minimisation, quality of products or services should not be compromised. Differentiation implies that one's products or services are unique, thus distinguishing oneself from other competitors. Adopting a focus strategy means that an organisation specifically attends to 'a particular buyer group, segment of the product line or geographic market' (38). Since 'focus' strategies can be incorporated when organisations opt to achieve either overall cost leadership or differentiation, thus arguably only two generic strategies exist in organisational practice: overall cost leadership and differentiation (Hampson and Kwok 1997).

The traditional design-then-build type of procurement has been used for decades and remains the primary project procurement approach (Rowlinson *et al.* 2008). The risks involved in fixed price contracts are substantial in highly competitive markets. This situation has further deteriorated as weather becomes more extreme and global financial markets become more fragile. Except for local housing markets, cost leadership cannot be achieved without undermining the performance. Since construction is a labour- intensive industry, shortage of labour is the prime factor driving wage increases. Sharing common inputs among competitors is a potential problem for any attempt to achieve overall cost leadership (Dess *et al.* 2012: 168). Cost differentiation is not easy to achieve.

Differentiation is another possible strategy whereby an organisation distinguishes itself from its competitors (Porter 1998: 37). Achieving differentiation creates competitive edge. Differentiation of product can be in the forms of prestige, technology, innovations, features, customer services and dealer network (37). Porter reminds us of the adverse effect of being 'stuck in the middle' (41), i.e., failing to achieve in at least one of the strategies. When an organisation develops a certain extent of overall cost leadership as well as differentiation, the organisation is likely to have compromised in profitability. Customers demanding low-cost products often only offer a low price. Furthermore, the organisation is likely to lose clients who look for differentiated product or services. It is therefore suggested that CCOs should avoid cost competition; instead differentiation by unique organisational strength is preferred. To achieve this, developing soft power is advocated.

Developing soft power for construction contracting organisations

Soft power refers to those non-technology-based competitive edges possessed by a CCO. Having distinctive soft power is a way to differentiate oneself. Change and innovation are regarded as main ingredients of competitive organisations (Creed 2012: 257). In fact, the World Economic Forum regards innovation as one of the pillars to evaluate competitiveness (WEF 2013). A CCO must possess the organisational capacity to innovate that can be translated as breaking out of conventional thinking and working. Organisational capacity for change is 'a dynamic, multidimensional capability that enables an organisation to upgrade or revise existing organisational competencies, while cultivating new edges that enable the organisation to prosper' (Judge 2011: 14).

Soft power is a political science concept and refers to the ability to get the desired outcome through attraction rather than coercion or payments (Nye 2004: 5). It involves assertion through persuasion. Under soft power theory, cultural and economic influence is the manifestation of soft power. Exploration of this form of power in construction offers renewed perspectives in analysing means to raise the competitiveness of CCOs. Enhanced knowledge in this regard will pave ways for more effective management; from both inter- and intra-organisational perspectives. CCOs equipped with substantive soft power should be able to cope with unanticipated challenges. Three forms of soft power are discussed: (1) ability to work collaboratively in a competitive environment; (2) ability to continue to improve as a learning organisation; and (3) ability to minimise non-productive use of scarce resources.

Ability to work collaboratively in a competitive environment

Construction contracts are inevitably incomplete due to the impossibility in foreseeing all the eventualities during the contract preparation stage. With rising project complexity and high degree of uncertainties (Mitropoulos and Howell 2001), construction contracts are unable to provide exhaustive measures to handle all eventualities. Incomplete contracts inform opportunism. CCOs may act opportunistically to recoup dissipating profit by lodging ill-founded claims ex post (CIRC 2001). Opportunistic behaviour is undesirable and fuels adversity among contracting parties. It has been advocated that contracting parties should work collaboratively to solve problems that may arise. Moreover, scepticism among contracting parties remain the norm in construction contracting business. To combat this, developing a trusting environment is considered to be conducive in engendering cooperation whereby the strength of construction contracting parties can be synergised (Gulati 1995; Janowicz and Noorderhaven 2006; Madhok 1995).

Trust has been defined as the willingness to subject to vulnerability (Mayer *et al.* 1995). This definition is widely adopted in psychology, organisational behaviour and strategic economics (Lewicki *et al.* 1998; Mishra and Spreitzer 1999; Bhattacharya *et al.* 1998). This assertion is based upon positive expectations and attitude towards the reciprocating behaviour of a counterpart (Rousseau *et al.* 1998). In inter-organisational context, the level of trust rises with multiple positive interactions among contracting parties. Trust can suppress opportunism that undermines cooperation. Therefore trust is an important element in the collaborative working environment. Trust is also a prerequisite to inter-organisational coordination (Asad *et al.* 2005; Tatum 1987) especially when sharing of sensitive information is necessary to deal with difficult and tricky construction problems. Chance of conflict escalation can be reduced if the contracting parties trust each other.

Collaborative working is more likely to engender synergistic use of resources. Regrettably, it is difficult to create a collaborative environment in construction contracting. This may be due to differences in aims and culture. For example, the aim of the project client is to receive a quality product within budget and time, while the aim of CCOs is maximising profit. Both make perfect sense from a business perspective. Furthermore, the level of collaboration between CCOs is not stable. It is therefore crucial for the CCOs to be cautious. The moves of the counterparts, either cooperative or aggressive in nature, are reflective of their underlying motive. Thus, one has to undertake a strategic decision-making process to decide whether he would act cooperatively or aggressively. As such, it is important to identify the drivers of cooperative moves or aggressive moves so that construction professionals can make good use of tactics to achieve desirable project outcomes. With collaborative attitudes, strategic alliances can be formed. Strategic alliances embrace genuine exchange of useful information with the aim of achieving overall gain of the alliance (Pearce and Robinson 2009: 361).

Having a sustainable collaborative attitude is a soft power. Ideally, this attitude should be part of the DNA of a forward-looking CCO. Organisational culture can be defined as:

> a pattern of basic assumptions – invented, discovered, or developed by a given group as it learns to cope with its problems of external adoption and internal integration – that has worked well enough to be considered valid and, therefore, to be taught to new members as the correct way to perceive, think, and feel in relation to those problems.
>
> (Judge 2011: 92)

Organisational culture is said to be a combination of characteristics of different groups and individuals, the working approaches of the organisational members, types of leaderships and politics (Creed 2012: P.206). Organisational culture is expressed through members' audible behaviour such as language norms and rituals (Creed 2012: 206; Judge 2011: 93). The audible behaviour is the manifestation of values and beliefs (Creed 2012: 206). Organisational culture is often recognised as 'social glue' (Judge 2011: 93). More importantly, organisational culture can serve as a device guiding the behaviour of members of an organisation and through which their behaviour can be shaped (Egbu *et al.* 1998). Thus, organisational culture can be a significant soft power of a CCO if it drives performance. More importantly, organisational culture is often difficult for competitors to imitate, unlike other resources such as human resources and technology (Judge 2011: 93). Human resources can be highly mobile. In 2013, the staff turnover rate in the construction/property development/real estate sector is the highest (45.5 per cent) among different fields in Hong Kong (HKIHRM 2013). Jochimsen and Napier commented 'an organisation with high staff turnover rate or that lacked stable membership or shared corporate history, may be unable to sustain a social understanding over time' (2013: 235). Shared corporate history underpins the culture of an organisation. It may not be too obscure for CCOs to be innovative if a collaborative attitude can be imagined.

An innovative culture is particularly important (Creed 2012: 222). Fostering an innovative culture involves encouragement of learning (ibid.).

It is impossible for an organisation to innovate without changes (Atkin 1999). Egbu *et al.* (1998: 605) stressed that innovation in construction 'is a complex social process which involves both the "soft" human side issues and the "hard" issues of tools and techniques' (Atkin 1999; Egbu *et al.* 1998; Blayse and Manley 2004). Thus culture can be a source of encouraging innovation. Nurturing an influential innovative culture can be viewed as a psychological process to motivate the members of an organisation to be innovative. Broadly, the necessary success factors include: support from top management, employees' involvement, communication and information exchange and sharing of risks and liabilities.

Support from top management

Changing corporate orientation is a trigger as well as a critical element to unleash potential in innovation. Initiating top management of an organisation can drive quick results as the corresponding organisational strategies can be promptly instigated. When the top management of an organisation is committed and supportive to innovation development, an innovative culture is likely to find seeds and germinate. Support from the top level of an organisation is a prerequisite for developing innovations. This requires vision from top management.

Communication and information exchange

Many innovations are solutions to problems encountered in daily operations. In other words, the inspirations are need-driven. Members of an innovative CCO are not content with traditional solutions. Instead, they would seek alternatives and in some circumstances find creative solutions. This resembles a bottom-up approach in driving innovations. When top management formulates strategic directions to encourage innovation and creativity, a top-down approach is used. To allow lateral communication, cross-functional teams can be established. Multidimensional flow of information and communication is allowed in this setting. Thus an interactive corporate management structure has been formed and this is conducive to innovative thinking in a holistic manner. The cohesion between the participants has been enhanced (Atkin 1999). On this note, Egbu *et al.* (1998) suggested that an environment promoting top-down; bottom-up and lateral communications is conducive for innovation. Involvement of all members in innovative practice can also encourage innovations (Judge 2011: 108).

Sharing of risks and liabilities

Developing innovations involves technological and financial risks (Egbu *et al.* 1998). An incentivising risk-sharing system is believed to encourage innovations across an organisation. A CCO can arrange competition and awards to support its members' innovative endeavours. The themes of the competitions can be technical, organisational and beyond. Through these awards, all levels of members are given the chance to participate in the competitions to stimulate innovation. In addition, monetary rewards can be given and the award will be presented in prize-giving ceremony. This arrangement serves as recognition and a token of appreciation to members' innovations. Section B of the book first discusses the relationship between organisational culture and performance, then the ways to engender trust in construction contracting is explored. The aggressive and cooperation drivers are analysed in the last chapter of Section B.

Ability to continue to improve as a learning organisation

Being able to adjust and improve when faced with challenges is essential for CCOs. Furthermore, if learning can be achieved through these adjustments, a CCO will raise its operation to the next higher level. Organisational learning therefore is a soft power that helps CCOs to improve efficiency and effectiveness. It is a process of forming and developing skills, knowledge, attitude and behaviour (Pettinger 2010: 99) from lessons of success and failure. Attitudes are the propositions adopted by individuals. Knowledge can be operating know-how, relationship with and knowledge of customer networks or technical knowledge upon which products or processes are based (Pearce and Robinson 2009: 364). Individual learning is different from organisational learning and learning by individuals does not guarantee an organisation can benefit (Castaneda and Rios 2007; Senge 2006: 129). Organisation can be conceptualised as 'the product of thoughts and action of members' (Castaneda and Rios 2007). Organisational learning requires a 'transference process of knowledge among people, with the purpose of institutionalisation' (Castaneda and Rios 2007; Wang and Ahmed 2003). Institutionalisation refers to a process of embedding learning by individuals and groups into organisations including system, structures, procedures and strategy (Castaneda and Rios 2007; Crossan *et al.* 1999). Learning also includes unlearning obsolete or old ways of thinking so that the capabilities of members of a CCO are kept updated and current. As such, learning can facilitate members of the construction contracting organisations to engage in productive and effective problem solving. Organisational learning arouses creativity and underpins innovations (Dess *et al.* 2012: 411).

To become a learning organisation, Senge (2006) has identified several core qualities: personal mastery, mental models, shared vision, team learning and system thinking. Personal mastery refers to motivation to learn and continuous learning. Training and persistent professional development can facilitate continuous learning (ibid.). Learning can be motivated by job promotion or salary increase. Mental models refer to decisions and behaviours that are influenced by perception. Mental models are manifested by institutional practices (ibid.). To institutionalise learning, sharing of knowledge among organisational members is made compulsory. Due to the project-based nature of construction contracting organisations, the learning between projects has been hindered (Blayse and Manley 2004). Knowledge codification is thus required to allow easier access of information and knowledge between members. This often requires a knowledge centre where the knowledge can be kept in a systematic manner. Workshops to gather practitioners and sharing groups can be formed for exchange of information and expertise. The cohesion between the participants has been enhanced (Atkin 1999). Through experience-sharing sessions, a learning environment among professionals from different disciplines has been created. Sharing sessions facilitate brainstorming and generation of ideas. Flow and exchange of knowledge can also be enhanced. Shared vision refers to the answer of the questions of an organisation, 'what we want to create' (Senge 2006). According to Senge (2006), a shared vision offers focus and energises learning. Shared vision is to be given by the top management of an organisation. Thus commitment and support from top management is required to effect learning. Team learning refers to 'process of aligning and developing the capacity of a team to create the results its members truly desire' (ibid.). It builds on shared vision and personal mastery. System thinking refers to 'a discipline for seeing wholes' (ibid.: 68). It requires the understanding of how the problem can be solved by taking into account the overall systems and its environment (Judge 2011: 66).

Learning involves behaviour modifications through positive reinforcement (Creed 2012: 38). When the behaviours of the members of a CCO are collectively shaped and moulded into a specific pattern through learning, the specific pattern is referred to as organisational culture (53). It is worth noting that organisational learning is closely related to organisational culture. Section C of this book examines in detail the aforementioned concept of learning and unlearning of CCOs.

Ability to minimise non-productive use of scarce resources

Protracted dispute without resolution consumes valuable resources unproductively. Negotiation is well-recognised as the most cost-effective way to resolve disputes and conflicts. Construction professionals have to negotiate with their counterparts as construction often involves coordination, interface, rights and obligations that may not be clearly defined. Negotiation involves the process of offering and demanding until both parties reach agreement. Moreover, negotiators are representing their own organisations' interest and have to make decisions whether to accept the offer from counterparts. In fact, not every negotiation would end with agreement or consensus. During negotiation, the tension and conflict may escalate and result in deadlock and withdrawal. Therefore, tensions between negotiating parties have to be monitored to prevent negotiators leaving the negotiating table. Negotiation failure is referred to as unproductive use of scarce resources because much more extra effort is required should formal litigation and arbitration be necessary.

In a transaction economics perspective, a rational negotiator would aim to maximise his/her profit. With profit maximisation in mind, perception of equity or fairness in negotiation becomes secondary. During negotiation, cautious negotiating parties tend to disclose as little information as possible. When a negotiator perceives unfairness, he/she may reciprocate with dishonest actions for his/her own profit. For example, negotiators could withhold information within permissible boundaries. Contractual rights take precedence over ethical considerations (Adnan *et al.* 2011). Relationships between the negotiation parties would be soured and non-conducive for settlement.

On the contrary, if win-win solutions are respected, the chance of reaching mutually beneficial terms saves the parties' energy and effort. Analysing equity sensitivity of members of a CCO is thus useful as this is indicative of the propensity of a negotiator in practising constructive negotiation. Accurate prediction of the behaviour of the negotiator allows management to assign a negotiator suitable and appropriate for the task. Furthermore, negotiation and organisational culture are closely related. Although different organisations have their unique culture, forming a culture that facilitates win-win situations can be common to progressive CCOs. It is worth observing the organisational culture of counterparts so that one may be able to interpret their negotiating styles. With this knowledge, appropriate matching strategies can be developed. Above all, having negotiated settlement is one of the essential means to minimise non-productive use of scarce resources. It is a soft power that a competitive CCO should possess. Section D of the book is devoted to the ways to achieve amicable negotiation outcome in construction contracting.

Summary

The construction industry is one of the pillars of many economies. With ever-increasing project complexity and escalating cost, competition among CCOs is fierce. Among the five competitive forces identified by Porter (1998), it is advocated that CCOs should differentiate themselves by developing unique competitive advantages. CCOs can achieve this by developing soft power, i.e., non-technology-based competitive edges. It is suggested that soft power of CCOs include three types of ability: (i) to work collaboratively in a competitive environment; (ii) to continue to improve as a learning organisation; and (iii) to minimise non-productive use of scarce resources.

Acknowledgements

The work described in this chapter is fully supported by a City University Strategic Research Grant (No. 7004036).

References

Adnan, H., Hashim, N., Yusuwan, N. M. and Ahmad, N. (2011) Ethical issues in the construction industry: Contractor's perspective, in *Proceedings of Asia Pacific International Conference on Environment-Behaviour Studies*, 7–9 December.

Asad, S., Fuller, P., Pan, W. and Dainty, A. R. J. (2005) Learning to innovate in construction: A case study, in F. Khosrowshahi (ed.), *Proceedings 21st Annual ARCOM Conference*, 7–9 September 2005, London. *Association of Researchers in Construction Management*, 2, 665–74.

Atkin, B. (1999) *Study of Innovation in the Construction Sector*. Brussels: European Council for Construction Research, Development and Innovation.

Bhattacharya, R., Devinney, T. M. and Pillutla, M. M. (1998) A formal model of trust based on outcomes. *Academy of Management Review*, 23(3), 459–72.

Blayse, A. M. and Manley, K. (2004) Key influences on construction innovation. *Construction Innovation: Information, Process, Management*, 4(3), 143–54.

Castaneda, D. I. and Rios, M. F. (2007) From individual learning to organizational learning. *The Electronic Journal of Knowledge Management*, 5(4), 363–72.

Census and Statistics Department (2014) *Report on the Quarterly Survey of Construction Output, First Quarter 2014*. Hong Kong: Hong Kong SAR Government.

Construction Industry Review Committee (CIRC) (ed.) (2001) *Construct for Excellence*. Hong Kong: Construction Industry Review Committee.

Creed, A. C. (2012) *Organisational Behaviour*. Oxford: Oxford University Press.

Crossan, M. M., Lane, H. W. and White, R. E. (1999) An organizational learning framework: From intuition to institution. *Academy of Management Review*, 24(3), 522–37.

Dess, G., Lumpkin, G. T., Eisner, A. B. and McNamara, G. (2012) *Strategic Management, Creating Competitive Advantages* (6th edn). Columbus, OH: McGraw Hill.

Egbu, C. O., Kaye, H. G. R, Quintas, P., Schumacher, T. R. and Young, B. A. (1998) Managing organizational innovations in construction, in W. Hughes (ed.), *Proceedings of 14th Annual ARCOM Conference*, University of Reading. *Association of Researchers in Construction Management*, 2, 605–14.

Fiol, C. and Lyles, M. (1985) Organizational learning. *Academy of Management Review*, 10(4), 803–13.

Gulati, R. (1995) Does familiarity breed trust? The implications of repeated ties for contractual choice in alliances. *The Academy of Management Journal*, 38(1), 85–112.

Hampson, K. D. and Kwok, T. (1997) Strategic alliances in building construction: a tender evaluation tool for the public sector. *Journal of Construction Procurement*, 3(1), 28–41.

Hong Kong Housing Authority (HKHA) (2011) Going an extra mile to innovate the procurement system, available at: www.housingauthority.gov.hk/en/about-us/publications-and-statistics/housing-dimensions/article/20110318/going-an-extra-mile-to-innovate-the-procurement-system.html

HKIHRM (2013) *HR Service Providers Directory 2013*. Hog Kong: Hong Kong Institute of Human Resources Management. www.hkihrm.org/index.php/component/content/article/569-s/mb/member/hr-statistics/manpower-trend-manpower-change-turnover-rate-vacancy-rate/1754-turnover-rate-and-vacancy-rate-2nd-half-of-2013-en

Janowicz, M. K. and Noorderhaven, N. G. (2006) Levels of inter-organizational trust: Conceptualization and measurement, in R. Bachmann and A. Zaheer (eds.), *Handbook of Trust Research*.Cheltenham: Edward Elgar, 264–79.

Jochimsen, B. and Napier, N. K. (2013) Organizational culture, performance, and competitive advantage: What next?, in V. R. Kannan (ed.), *Strategic Management in the 21st Century, Volume 2: Corporate Strategy*. Santa Barbara, CA: Praeger, 233–54.

Judge, W. Q. Jr. (2011) *Building Organizational Capacity for Change: The Strategic Leader's New Mandate*. New York: Business Expert Press.

Lewicki, R., McAllister, D. J. and Bies, R. J. (1998) Trust and distrust: new relationships and realities. *Academy of Management Review*, 23(3), 438–58.

Madhok, A. (1995) Revisiting multinational firms' tolerance for joint ventures: A trust-based approach. *Journal of International Business Studies*, 26(1), 117–37.

Mayer, R. C., Davis, J. H. and Schoorman, F. D. (1995) An integrative model of organizational trust. *Academy of Management Review*, 20(3), 709–34.

Mishra, A. K. and Spreitzer, G. M. (1999) Giving up control without losing control: Trust and its substitutes' effects on managers' involving employees in decision making. *Group and Organization Management*, 24(2), 155–87.

Mitropoulos, P. and Howell, G. (2001) Model for understanding, preventing and resolving project disputes. *Journal of Construction Engineering and Management*, 127(3), 223–31.

Nye, J. S. (2004) *Soft Power: The Means to Success in World Politics* (1st edn). New York: Public Affairs.

Pearce, J. A. and Robinson, R. B. Jr. (2009) *Formulation, Implementation and Control of Competitive Strategy* (11th edn). Irwin: McGraw Hill.

Pettinger, R. (2010) *Organisational Behaviour, Performance Management in Practice*. New York: Routledge.

Porter, M. E. (1998) *Competitive Strategy, Techniques for Analyzing Industries and Competitors*. New York: The Free Press.

Rousseau, D., Sitkin, S. B., Burt, R. S. and Camerer, C. (1998) Not so different after all: a cross-discipline view of trust. *Academy of Management Review*, 23(3), 393–404.

Rowlinson, S., Tas, Y. K. and Tuuli, M. M. (2008) A cultural perspective on stakeholder management in the Hong Kong construction industry, in *Proceeding of International Conference on Multi-national Construction projects "Securing high Performance through Cultural awareness and Dispute Avoidance"*, Shanghai, China.

Senge, P. M. (2006) *The Fifth Discipline: The Art and Practice of the Learning Organization*. New York: Doubleday/Currency.

Tatum, C. B. (1987) Process of innovation in construction firm. *Journal of Construction Engineering and Management*, 113(4), 648–63.

The Government of the HKSAR (HKSARG) (2002) Tender evaluation of works contracts, Technical Circular (Works) No. 22/2002.

The Government of the HKSAR (HKSARG) (2004) Tender evaluation of works contracts, Technical Circular (Works) No. 08/2004.

The Government of the HKSAR(HKSARG) (1998) Tender evaluation of works contracts, Technical Circular (Works) No. 23/98.

Wang, C. and Ahmed, P. (2003) Organizational learning: a critical review. *The Learning Organization*, 10(1), 8–17.

World Economic Forum (WEF) (2013) *The Global Competitiveness Index Report 2013–2014*. Geneva.

Section B

Collaborative working in a competitive environment

2 Exploring organisational culture–performance relationship in construction

Sai On Cheung, Peter Shek Pui Wong and Wai Yin Wu

Introduction

Construction aspires to be an efficient industry (CIRC 2001; Egan 1998). Some of the well-known barriers to this aspiration include, inter alia, confrontational contracting behaviour, lack of innovation and indifference to research and development (Egan 1998; Zeng *et al.* 2009). Some notable successful construction contracting organisations have insightfully departed from the conventional 'construction only' business model. These organisations have in fact realised the potential markets by departing from entrenched practices. This move involves a form of cultural change and can be risky. However, changes have brought life to these organisations by widening their scope of operation. In fact, organisational culture (OC) has been identified as one of the essential factors that affect the efficiency and productivity of a firm (Alas *et al.* 2009). It has been strongly advocated that through cultivating and maintaining an induce culture which is performance stimulating, the efficiency of construction contracting organisations can be improved (Gordon and DiTomaso 1992). Successful project delivery relies on the concerted effort of all participants in realising designs into physical objects. The impact of their performance on projects is particularly apparent because of their front-line positions. In this regard, this study aims to investigate the aspects of organisational culture that empower a construction contracting organisation. The study is organised as follows. First, the impact of organisational culture on performance is reviewed. Second, identifiers of OC and performance indicators of construction contracting organisations are listed. Third, an OC and construction contracting organisations' performance relationship framework is proposed. Fourth, an empirical test of this OC-performance relationship framework is presented. Finally, the findings and their implications on project management are discussed.

Organisational culture and performance

The concept of culture has become a major theme of organisational studies since the mid 1980s. One of the influential works is the Organisational

Culture and Leadership model developed by Schein (1985). Organisational culture is defined as a pattern of basic assumptions – invented, discovered, or developed by a given group as it learns to cope with its problems of external adaptation and internal integration – that has worked well enough to be considered valid and, therefore, to be taught to new members as the correct way to perceive, think and feel in relation to those problems (ibid.). This definition reveals, by its focus on assumptions, that when considering culture we are dealing with implicit assumptions as much as explicit or overt behaviour. A 'strong' culture is one where the implicit and explicit assumptions are in harmony and are deeply entrenched and change-resistant. Similarly, Cole (1997) considered culture as a two-tiered set of 'shared values, norms and beliefs within an organisation'. On the surface it is the explicit culture, which manifests itself in the 'official' organisational structure and communications. Beneath the surface it is the implicit culture, which management and staff consider really important. Moreover, Smircich (1983) defined OC as the social glue that holds together members in an organisation. It expresses the social ideals, values and beliefs that members of an organisation come to share. These values or patterns of belief are manifested in symbolic devices such as myths, rituals, stories, legends and specialised language (ibid.). Schein (1985) also pointed out that there may be several cultures operating within an organisation: a managerial culture, various occupationally-based cultures in functional units, group cultures based on geographical proximity and worker cultures based on shared hierarchical experiences. The organisation as a whole will have an overall culture 'if that whole organisation has a significant shared history'. Organisational culture therefore has several important functions. First, it conveys a sense of identity for organisation members. Second, it facilitates the generation of commitment to something larger than the self. Third, it enhances system stability. And fourth, organisational culture serves as a sense-making device that can guide and shape members' behaviour (Alas *et al.* 2009; Hofstede 2001; Peters and Waterman 1982). Taking into account contextual factors, positive and enduring OC therefore can have a positive effect on individual and organisational performance (Denison 1990). With these propositions, research into the relationship between organisational culture and performance has surged (Ankrah and Langford 2005; Cheung *et al.* 2010; Denison 1990).

Employing the Organisational Culture Assessment Instrument (OCAI) developed by Cameron and Quinn (1999), Zhang and Liu (2006) proposed a behaviour–outcome model in analysing the organisational culture profiles of construction enterprises in China. Based on a case study of two Dutch contractors, Caerteling *et al.* (2006) found that dynamic, innovation-oriented contractors adopt more progressive policies in conducting their business. The study by Cheng and Liu (2007) also identified a significant correlation between the clan culture of construction firms and success in implementing Total Quality Management. More recently, Ozorhon *et al.*

(2008) examined the extent to which the performance of an international joint venture (IJV) is affected by the organisational culture of the collaborating firms. These studies suggested that performance improvement in an organisation is a result of successfully translating values and beliefs into policies and practices. Although the OC assessment tools adopted in the above studies were not construction specific, the findings indicate a close relationship between OC and performance. Nonetheless, the nature of such a relationship remains unanswered in the construction context and this deserves further investigation. The study reported in this chapter aims to fill this research gap. For this study, the hypothesis therefore is:

H1: *The performance of a construction contracting organisation is positively affected by its organisational culture.*

To test this hypothesis, a framework describing the relationship between identifiers of organisational culture and performance indicators is analysed using Structural Equation Modelling.

Understanding organisational culture

One of the early influential studies on organisational culture was on its definition and implications for managers. Schein (1986) defined 'culture' as a pattern of basic assumptions – invented, discovered, or developed by a given group as it learns to cope with its problems of external adaptation and internal integration. Organisational culture thus serves the leader of an organisation through nurturing the value system created by him to both serving and incoming members. According to Schein, a strong culture is therefore one where the implicit and explicit assumptions are in harmony. Schein (ibid.) further pointed out that there may be several cultures operating within an organisation: managerial culture that is occupationally based, group culture that is derived from geographical proximity and worker culture that is based on shared hierarchical experiences. The organisation as a whole will have an overall culture 'if that whole organisation has a significant shared history'. Furthermore, Cole (1997) considered culture as a two-tiered set of 'shared values, norms and beliefs within an organisation'. On the surface is the explicit culture, which manifests itself as the 'official' organisational and communication structure. Beneath the surface is the implicit culture that management and staff consider of real importance. Cole believes that implicit culture is probably closer to reality. This conception is similar to the 3-level framework of organisational culture proposed by Schein (2004) as shown in Figure 2.1.

Levels of culture can be analysed by the degree of visibility (Schein 2004). Artefacts are at the base level and include all the phenomena that one sees, hears and feels in encountering a new group with an unfamiliar culture. Artefacts can be observed but it is not easy to apprehend the deeper

Levels of culture	Examples	Characteristics
Artefacts	Visible organisational structures and processes	Surface but hard to decipher
Espoused beliefs and values	Strategies, goals, philosophies	Sharing with espoused justification
Underlying assumptions	Unconscious, taken-for-granted beliefs, perceptions, thoughts and feelings	Concordance for being the ultimate source of values and action

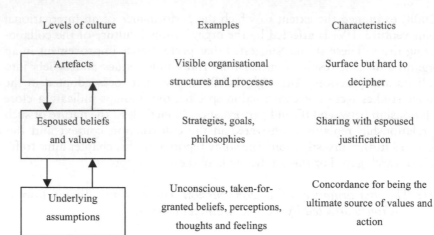

Figure 2.1 Organisational culture framework (adapted from Schein 2004)

assumptions per se. Nonetheless, these artefacts reflect the beliefs and values shared by members of an organisation. The innermost level of culture is the basic assumptions to which members of an organisation subscribe. It represents a level of concordance in the basic assumptions that are believed to be non-confrontable and non-debatable. In this regard, Schein (2004) advocates that leadership and organisational culture are the two sides of a coin because very often these basic assumptions are those of the leader.

From a project management perspective, Parker and Skitmore (2005) found that dissatisfaction with organisational culture is the primary reason causing staff turnover. Adenfelt and Lagerström (2006) found that organisational culture is the most prominent enabler in enhancing knowledge management in transnational projects. Atkinson *et al.* (2006) advocated that management of uncertainty is a necessary condition for effective project management. In this regard, it is suggested that organisations need to pay attention to culture and develop appropriate infrastructure and capability to manage uncertainties. Marrewijk (2007) highlights the danger of culture mismatch when the business model of an organisation changes. Likewise, similar project management issues will arise when a project manager needs to manage a group of organisations displaying different cultural characteristics (Patanakul and Milosevic 2009). These studies also bring out the need to consider the contextual factors in organisational culture studies (Saffold 1988). Two interesting study gaps can be noted. First, the concept of organisational culture in construction project management has only been explored under specific context. For example, the studies of Parker and Skitmore (2005), Adenfelt and Lagerström (2006) and

Atkinson *et al.* (2006) address staff turnover, knowledge and risk management issues respectively. In this regard, organisational culture in construction should be more holistically defined. Second, are there significant differences in views on organisational culture among major groups of stakeholders. In this study, three groups of stakeholders are considered: developers, consultancy offices and contractors. The study reported in this chapter aims to contribute to these two gaps.

To accomplish this objective, a literature review was performed to long-list artefacts that identify organisational culture in construction. With these, a questionnaire was designed to collect the degree of agreement of construction professionals in Hong Kong on the use of the long-listed artefacts as organisational culture identifiers in construction. Principal Component Factor Analysis (PCFA) was then applied to put artefacts of similar nature into organisational culture factors. The relative importance rankings of these factors were then assessed. Referring to the significance rankings, management shall have a better understanding about the factors shaping construction contracting organisations' behaviour. Efforts to foster organisational culture conducive to the achievement of project goals can be more focused by setting targets, directing resources and establishing benchmarks for the respective artefacts. In other words, monitoring the wellbeing of the artefacts shall inform the status of organisational culture as well as management action.

Identifiers of organisational culture

Organisational culture implies the existence of certain closely associated and interdependent dimensions (Peters and Waterman 1982). Ankrah and Langford (2005) preferred the use of dimensions to describe OC because of its flexibility to reflect the values being assessed. A list of dimensions can be compiled to reflect the various perspectives adopted by different researchers. However, having too many dimensions may make the framework ineffective as attention may be diverted to the fine and subtle differences among dimensions (Hofstede 2001). The identifiers of organisational culture suggested by different researchers are summarised in Table 2.1. It is proposed that organisational culture can be identified by eight cultural dimensions: goal clarity (GC), coordination and integration (CI), conflict resolution (CR), employee participation (EP), innovation orientation (IO), performance emphasis (PE), reward orientation (RO) and team orientation (TO).

For the purpose of this study, it is also necessary to devise operation statements to characterise the respective OC dimensions. These become the measurement statements to be used to collect data for the testing of the relationship framework. This methodology has been successfully used by Butler (1991) in measuring trust between managers and subordinates, and by Cummings and Bromiley (1996) in measuring trust between units of an organisation.

Table 2.1 Identifiers of organisational culture (OC)

Organisational culture (OC) identifiers	OC operational statements	Mean score	A	B	C	D	E	F	G
Goal clarity (GC)	How well the employees know what they need to do to succeed in the long run (GC1)	5.12	*	*			*	*	*
	The extent to which organisation's goals are reasonably and clearly set with regular reviews (GC2)	4.91	*	*			*	*	*
	The extent to which employees' effort is directed to accomplish the organisation's goal (GC3)	4.73	*	*			*	*	*
Coordination and integration (CI)	The effectiveness of resolving problems between departments (CI1)	4.34	*	*	*	*	*	*	
	The extent to which sharing of information between departments is encouraged (CI2)	4.79	*	*	*	*	*	*	
	The extent to which cooperation and assistance across departments is encouraged (CI3)	4.69	*	*	*	*	*	*	
Conflict resolution (CR)	The extent to which the employees accept criticism or negative feedback without becoming defensive (CR1)	4.27			*				*
	The extent to which employees are encouraged to share the responsibility of things that go wrong in their work group (CR2)	4.42			*				*
	The atmosphere of trust in this organisation (CR3)	4.40			*				*
Employee participation (EP)	The extent to which employees are encouraged to have some input on decisions that affect their work (EP1)	4.47	*	*			*	*	*
	The extent to which organisations allow employees to participate in the decision-making process (EP2)	4.40	*	*			*	*	*
	The extent to which employees are consulted in respect of decisions regarding what the organisation plans to do (EP3)	3.99	*	*			*	*	*
Innovation orientation (IO)	The extent to which the organisation helps employees to obtain the resources necessary to implement their innovative (IO1)	4.58	*		*				*
	The extent to which employees are encouraged to search for better ways of getting the job done (IO2)	4.53	*		*			*	*

Table 2.1 continued

Organisational culture (OC) identifiers	OC operational statements	Mean score	References						
			A	B	C	D	E	F	G
	The extent to which employees are encouraged to be creative and innovative (IO3)	4.55	*						*
	The willingness of the organisation to take reasonable risk in response to changes of business environment. (IO4)	4.35	*						*
Performance emphasis (PE)	The extent to which employees are coached to improve their skills so they can achieve higher levels of performance (PE1)	4.71	*	*	*		*	*	*
	The establishment of a set of performance standards for employees (PE2)	4.45	*	*	*		*	*	*
	The extent to which the organisation emphasises delivering products with good quality (PE3)	5.14	*	*	*		*	*	*
Reward orientation (RO)	The extent of equitable rewards (RO1)	4.32		*			*	*	*
	The level at which performance appraisals are used as the basis to reward employees (RO2)	4.29		*			*	*	*
	The level at which emphasis is placed on rewarding employees for success rather than punishing them for failure (RO3)	4.51		*			*	*	*
	The extent to which the employees are adequately recognised and rewarded (RO4)	4.46		*			*	*	*
Team orientation (TO)	The extent to which the organisation emphasises team contributions rather then individual contributions (TO1)	4.74			*	*	*		*
	The extent to which the organisation emphasises building cohesive, committed teams of people (TO2)	4.58			*	*	*		*
	The extent to which members work as a team and exchange opinions and ideas (TO3)	4.35			*	*	*		*

References:
(A) Peters and Waterman 1982, (B) Cameron and Quinn 1999, (C) Denison 1990, (D) Hofstede 2001, (E) Ankrah and Langford 2005, (F) Zhang and Liu 2006, (G) Cheung *et al.* 2010

Performance indicators of construction contracting organisations

The use of indicators to evaluate organisational performance is very common (Xiao and Proverbs 2003). In construction, compliance with predetermined criteria regarding time, cost and quality are the key indicators typically used (ibid.). The KPI Working Group in the United Kingdom developed key performance indicators (KPI), which is one of the most popular performance evaluation frameworks in use. Under KPI, construction contracting organisations' performance is evaluated along a number of dimensions, including:

1 profitability;
2 productivity;
3 return on capital employed;
4 return on value added;
5 interest cover;
6 return on investment;
7 ratio of value added;
8 repeat business;
9 outstanding money;
10 time taken to reach final account.

Notwithstanding its popularity, KPI seems to be more appropriate in assessing performance at project level (Kagioglou *et al.* 2001). Furthermore, performance should not only be singularly assessed by the achievement of measurable benefits, but also by the effectiveness of contractors in sustaining performance improvements, such as their competence in addressing risk and its consequences (Law and Chuah 2004), learning from experience (Wong *et al.* 2008) and generating innovative ideas (Kagioglou *et al.* 2001). To this end, Bayliss *et al.* (2004) critically compared the strengths and weaknesses of a number of performance measurement systems in construction, including the KPI, the European Foundation for Quality Management Excellence Model (EFQM) and the Balanced Scorecard (BSC). They suggested that the BSC framework developed by Kaplan and Norton (1992) provides a more holistic assessment of performance that goes beyond the project level. Mohamed (2003) indicated the word 'Balanced' in BSC represents equal emphasis on both tangible and intangible elements representing the core values of the company. The 'Scorecard' records results systematically and indicates the successfulness in adopting appropriate strategies to address short- and long-term goals (Amaratunga *et al.* 2001). The BSC framework also enables the company's core values and strategies to be articulated and linked (Mohamed 2003). With reference to the work of Kaplan and Norton (1992), four strategic dimensions can be used to measure organisational performance: financial, customer, internal business process, and learning and growth. In construction, a number of studies have used the

Balanced Scorecard (BSC) framework to evaluate construction contracting organisations' performance (Kagioglou *et al.* 2001; Mohamed 2003). For example, Mohamed (2003) applied the BSC approach to develop construction safety performance indicators. Four strategic dimensions for evaluating organisations' construction safety performance were proposed: (1) management, (2) operation, (3) customer and (4) learning. Based on case studies conducted in the United Kingdom, Kagioglou *et al.* (2001) proposed that construction project performance should be evaluated under the headings of financial, internal business processes and customer perspectives. To embrace the holistic approach of the BSC framework, the 'Innovation and Learning' dimension is also included. As a result, for this study, the four key performance indicators of construction contracting organisations used are: financial (FIN), internal business processes (IBP), customer (CUS) and innovation and learning (INL). The operation statements of these indicators are listed in Table 2.2.

A framework relating organisational culture and performance

Based on the discussion in the foregoing two sections, an OC-performance relationship framework arranged in a Structural Equation Modelling format is proposed (Figure 2.2). The arrows in the figure represent the direction of the hypothesised influence. For example, GC can be identified by the attribute: 'how well the employees know what they need to do to succeed in the long run' (GC1). Hence an arrow extends from 'GC' to 'GC1'.

Structural Equation Modelling is used to investigate : (i) the relationship between OC identifiers and their operators; (ii) the relationship between performance indicators and their operators; and (iii) the relationship between OC and performance. As such, two stages of data analysis were performed. The first stage involved the checking of construct reliability and inter-relationships. The checking of construct reliability is done to validate the reliability of representing a latent variable by its observed variables (also called internal consistency). This check can be done by conducting Cronbach alpha reliability testing. The alpha ranges from 0 to 1. The higher the alpha value, the greater is the internal consistency of the construct. A value from 0.6 to 0.7 is regarded as 'sufficient' and a value greater than 0.7 is regarded as 'good' (Sharma 1996). The construct inter-relationships were then checked using Pearson correlation analysis with the aim of validating the proposed inter-relationships among constructs. Both the Cronbach alpha reliability testing and the Pearson correlation analysis were done using Statistical Package for Social Science (SPSS).

The second stage involved analysing the overall fitness of the model by investigating the fitness of the hypothesised relationships using Structural Equation Modelling (SEM) (Jöreskog and Sörbom 1996). SEM is a useful tool in theory development because it allows the researcher to propose and

Table 2.2 Performance indicators of construction contracting organisations

Performance indicators	Performance operational statements	Mean scores	References H	I	J
Financial (FIN)	Meeting predetermined goals on profitability (FIN1)	4.63	*	*	
	Meeting predetermined goals on revenue growth (FIN2)	4.68	*	*	
	Maintaining competitiveness in the market (FIN3)	4.88	*	*	
	Meeting predetermined goals on increasing shareholder returns (FIN4)	4.73	*	*	
Internal business processes (IBP)	Meeting predetermined goals on quality level (IBP1)	4.84	*	*	
	Meeting predetermined goals on cost control (IBP2)	4.65	*	*	
	Enhancing competence in identifying company's goals (IBP3)	4.71	*	*	
	Enhancing competence in maintaining the process of achieving the predetermined goals (IBP4)	4.75	*	*	
Customer (CUS)	Obtaining feedback from customers (CUS1)	4.86	*	*	
	Enhancing competence in satisfying customer's needs (CUS2)	4.76	*	*	
	Enhancing competence in keeping existing customers (CUS3)	4.90	*	*	
	Meeting predetermined goals on company vision about customer service (CUS4)	4.67	*	*	
Innovation and learning (INL)	Providing adequate training to employees (INL1)	4.85	*		*
	Providing adequate review of practice to adapt to market change (INL2)	4.72	*		*
	Enhancing competence in driving innovative ideas from employees (INL3)	4.49			*
	Enhancing competence in transforming employees' innovative ideas to decisions (INL4)	4.40			*

References:
(H) Kaplan and Norton 1992, (I) Kagioglou *et al.* 2001, (J) Kululanga *et al.* 2001

subsequently test propositions about the interrelationships among variables in a multivariate setting (Hair *et al.* 1998).

SEM integrates the analytical functions of both multiple regression analysis (MRA) and confirmatory factor analysis (Arbuckle and Wothke 1999; Molenaar *et al.* 2000). SEM can be used to represent, estimate and validate linear relations among observable and latent variables of a hypothesised network (Molenaar *et al.* 2000). Hair *et al.* (1998) described SEM as a multivariate technique for estimating a series of inter-related and inter-dependent relationships simultaneously. Molenaar *et al.* (2000) emphasised

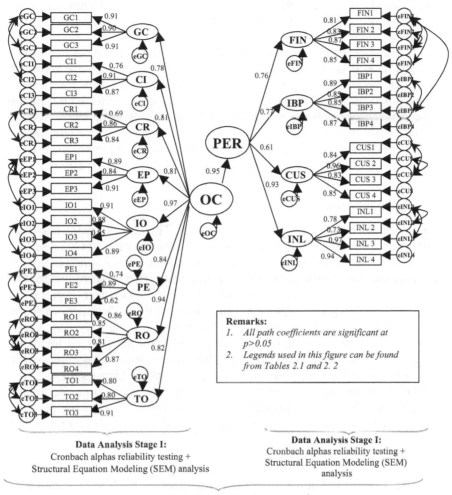

Figure 2.2 Final SEM of the OC-performance relationship framework

that the use of SEM can reduce the shortcomings of the MRA because the technique also accounts for errors in measurement when a large number of variables is involved. Thus, a more accurate representation of the overall results can be obtained from an SEM framework (Arbuckle and Wothke 1999). The software 'Analysis of Moment Structures 5.0' (AMOS) was used for the SEM analysis. The fitness of the relationship frameworks and the overall structural model was assessed using four Goodness of Fit (GOF) indices available from AMOS: root means square error of approximation (RMSEA), goodness-of-fit index (GFI), Tucker-Lewis index (TLI) and normal fit index (NFI).

The recommended acceptance thresholds of the GOF indices are shown in Table 2.3. If these are not achieved, model refinements are required. However, model refinements must be done carefully and with sound theoretical justifications. Thus a modification should only be effected if it makes good sense theoretically or practically (Arbuckle and Wothke 1999). AMOS offers modification suggestions for GOF value improvements. The suggestions include revising the relationship paths and adding covariance error paths between observed and latent variables (Molenaar *et al.* 2000).

Testing the frameworks

A questionnaire was used for data collection. The questionnaire has three parts: Part 1 – personal information; Part 2 – identification of OC; and Part 3 – performance assessment. The questions in Part 1 were aimed at soliciting personal information from the respondents. Part 2 includes 26 OC operational statements summarised in Table 2.1. Respondents were asked to express their degree of agreement on the use of these identifiers to represent the organisational culture of their companies, using a 7-point Likert scale. Part 3 includes 16 questions developed from the four performance indicators as summarised in Table 2.2. The respondents were requested to evaluate their company's performance using a 7-point Likert scale.

The questionnaire was sent to construction contracting organisations in Hong Kong. The target respondents include directors, project managers and professional-grade staff (including engineers and surveyors). A total of 185

Table 2.3 Goodness-of-fit (GOF) measures recommended levels and results (table format adapted from Molenaar *et al.* 2000)

Goodness-of-fit (GOF) measure	Recommended acceptance thresholds of the GOF indices	Model's GOF results		
		Stage IA: OC	Stage IB: performance	Stage II: OC-performance
Goodness-of-fit index (GFI)	0 (No fit) to 1 (Perfect fit)	0.84	0.86	0.82
RMSEA	<0.05 indicates very good fit – threshold level is 0.10	0.04	0.07	0.07
Tucker-Lewis index (TLI)	0 (No fit) to 1 (Perfect fit)	0.98	0.97	0.89
Normal fit index (NFI)	0 (No fit) to 1 (Perfect fit)	0.90	0.92	0.88

questionnaires were sent and 109 were completed and returned by the respondents. Six replies were excluded for incompleteness. As a result, 103 valid responses, representing a response rate of 55.7 per cent, were used for the analysis. Among the valid responses, over 70 per cent of the respondents had over 10 years working experience.

Findings and discussions

Descriptive statistics

With reference to Tables 2.1 and 2.2, it can be seen that the mean scores for the OC statements were higher than the mid-point of the 7-point scale (i.e. 3.5). This suggests that the respondents agreed that the eight OC identifiers and the respective operational statements can be used to describe the OC of their organisations. Furthermore, the mean scores for the performance operational statements ranged from 4.40 to 4.90 suggesting that the respondents generally rated their company's performance being above average.

Stage I: Testing the validity of the OC and performance constructs

To ensure the appropriateness of groupings of the operational statements for organisational culture and performance indicators, internal consistency of the constructs was checked. Table 2.4 details the results of the Cronbach alpha reliability tests. All groupings had Cronbach alpha values above 0.8, suggesting that the operational statements are significantly related to the respective OC identifiers and performance indicators. Hence, all the operational statements and their respective constructs are retained.

Table 2.4 Results of Cronbach alpha reliability testing

OC	Cronbach alpha value
Goal clarity (GC)	0.90
Coordination and integration (CI)	0.88
Conflict resolution (CR)	0.84
Employee participation (EP)	0.89
Innovation orientation (IO)	0.93
Performance emphasis (PE)	0.81
Reward orientation (RO)	0.92
Team orientation (TO)	0.880

Performance	Cronbach alpha value
Financial (FIN)	0.91
Internal business processes (IBP)	0.92
Customer (CUS)	0.93
Innovation and learning (INL)	0.91

The validity of the OC and performance constructs was then tested by SEM analyses. Considering the modifications suggested by AMOS, if necessary, refinements of the models can be made. No elimination of statements or constructs was suggested. This indicates that the constructs and relationship paths proposed in the conceptual model generally passed the statistical validity test. The suggested changes involved adding correlation paths between the error terms of the operational statements within the same construct (Arbuckle and Wothke 1999). The GOF indices for the final OC and performance constructs were satisfactory (results refer to Table 2.3).

Stage II: The OC–performance relationship framework

At this stage of the analysis, the refined constructs of OC and performance were combined to form the OC–performance relationship structural equation model. The validity of the structural model was then assessed as described earlier. Model refinements were performed until all GOF measures achieved the recommended levels (Molenaar *et al.* 2000; Wong and Cheung 2005). The set of model fit parameter values of the final SEM were: RMSEA = 0.07; GFI = 0.82; TLI = 0.89; and NFI = 0.88. The standardised regression weights and the GOF indices of the final SEM are presented in Figure 2.1 and Table 2.3 respectively.

To summarise, all the relationship paths as specified in the OC–performance SEM (refer to Figure 2.1) were found to be positive and significant at $p<0.05$. The SEM results suggest that:

1   OC can be represented by the eight OC identifiers as shown in Table 2.1: goal clarity (GC) (standardised regression weight = 0.77); coordination and integration (CI) (standardised regression weight = 0.78); conflict resolution (CR) (standardised regression weight = 0.81); employee participation (EP) (standardised regression weight = 0.81); innovation orientation (IO) (standardised regression weight = 0.97); performance emphasis (PE) (standardised regression weight = 0.84); reward orientation (RO) (standardised regression weight = 0.94); and team orientation (TO) (standardised regression weight = 0.82).
2   Construction contracting organisations' performance can be evaluated using the four strategic dimensions in the BSC framework: financial (FIN) (standardised regression weight = 0.76); internal business processes (IBP) (standardised regression weight = 0.77); customer (CUS) (standardised regression weight = 0.61); and innovation and learning (INL) (standardised regression weight = 0.93).
3   The effect of OC on performance (standardised regression weight = 0.95) is positive and significant at $p<0.05$. The hypothesis of this study is supported.

Discussions

The results of the study provide empirical support to the hypothesis that the performance of construction contracting organisations is positively affected by their organisational cultures. With reference to the standardised regression weights (SRW), innovation orientation (IO) and reward orientation (RO) are the highest among the eight OC identifiers. In fact, there is a clear gap between these two SRWs and the other six SRWs. Similarly, in the performance loop, the SRW of innovation and learning (INL) was 0.93 and significantly higher than the next highest of 0.77 for internal business process (IBP). The aforementioned findings suggest that innovation is the most distinct success factor in terms of both OC and performance.

Positive organisational cultures are those that enable the organisation to improve (Alas *et al.* 2009). Valuable culture is also often unique to the organisation and difficult to imitate (Barney 1986). Among the eight OC identifiers, IO and RO can be considered as positive cultural factors as suggested by Barney. Globalisation has revolutionised the construction market, changing it from a local industry to one involving international competition, especially for mega projects (Ankrah and Langford 2005). Hartmann (2006a, 2006b) considers an organisation that is able to consistently and profitably deliver better services than its competitors as having a real survival edge in the fierce global market. To achieve this, construction contracting organisations should develop cultures that can motivate and foster innovative behaviour amongst its members. Hartmann (2006a) identified organisational culture as an instrumental vehicle in driving innovation. The pivotal role of innovation in today's business can be evidenced by the success of companies such as Google and Apple. New ideas should not be lost in the organisations hierarchy, nor should creativity be undermined by daily routines. To these ends, a shared value of treasuring innovation within an organisation is the survival kit in today's competitive business environment. Construction contracting organisations should also develop a culture of rewarding employees and accepting innovative ideas in order to sustain their performance and competitive advantages. Based on a case study conducted in Switzerland, Hartmann (2006a) identified three managerial actions that construction contracting organisations can take to maintain staff involvement in and dedication to innovation: (1) establishing reward and incentive schemes to recognise innovative staff ideas; (2) allowing staff to take reasonable risks for implementing innovative ideas in operations; and (3) providing prompt and positive feedback on the staff proposals for innovative activities. Item (3) neatly brings our discussion to innovation and learning (IL) as a critical performance evaluator. Organisational learning embraces the broad concept of how organisations can learn from their own mistakes and experience in order to improve performance. Wong *et al.* (2008) identified that double-loop learning is a more effective learning style in terms of performance improvement. It was further suggested that performance-monitoring systems should be designed to facilitate learning. Further discussion on organisational learning is included in Section C of this book.

This study hypothesised that the performance of a construction contracting organisation is positively affected by its organisational culture. The final SEM suggests that the structural path from OC to PERF (regression weight = 0.95) was positive and significant as hypothesised (Ankrah and Langford 2005; Liu 1999). Innovation has been singled out as the key cultural factor in terms of driving performance, and is measured by the ability to innovate. Hammer (2004) illustrated how operational innovation allowed an automobile industry to survive and grow in a highly competitive and volatile automobile insurance market. Transformation was underpinned by a belief in keeping the customer satisfied. Operational procedures were constantly reviewed and adjusted to achieve customer satisfaction. Although construction is often identified as a 'conventional' industry that lacks the motivation to innovate, Winch (2003) aptly points out that this inefficiency label is the result of the narrow identification of construction. Taking the construction supply chain as a whole, innovation is occurring no less than in its manufacturing counterpart. Sexton and Barrett (2003) echoed this viewpoint and further suggested that small construction companies can contribute to innovation waves in their respective expertise. The findings of this study add strength to the general belief that cultural factors are having more profound effects on the long-term wellbeing of an organisation.

Summary

Construction aspires to be an efficient industry. A number of industry-wide studies have suggested the need to cultivate a culture that is more conducive to business success in construction. This study investigated the relationships between organisational culture and the performance of construction contracting organisations. Construction contracting organisations have been chosen for this study because they are at the production front-line in terms of realising designs into physical objects. Their performance thus has a direct impact on the intended output. OC identifiers and performance indicators were shortlisted from a literature review. These were then translated into operational statements. With data collected from Hong Kong contractors, the OC and performance constructs were first tested for internal consistency. These two constructs were then linked to form the OC–performance relationship framework that was tested using Structural Equation Modelling. The SEM results support the use of the eight OC identifiers for framing OC in construction and the four performance indicators for evaluating performance. The final OC–performance relationship framework indicates a positive relationship between organisation culture and performance. Innovation was found to be the most pivotal cultural factor, apparently for its decisive role in supporting creativity. To this end, proper recognition of, and reward for, innovation should be an integral part of performance evaluation. Notwithstanding the wealth of studies in organisational culture, with the respondents being members of construction contracting organisations, this study provides an extended perspective on the impact of project-based organisational culture on the performance of construction contracting organisations.

Acknowledgements

Special thanks to Miss Anna Ling Lam for collecting data for the study. The content of this chapter has been published in Volume 13(4) of the *Journal of Business Economics and Management* and Volume 29(1) of *the International Journal of Project Management*. These are used with the permission from Taylor & Francis and Elsevier respectively.

References

Adenfelt, M. and Lagerström, K. (2006) Enabling knowledge creation and sharing in transnational projects. *International Journal of Project Management*, 24, 191–8.

Alas, R., Kraus, A. and Niglas, K. (2009) Manufacturing strategies and choices in culture contexts. *Journal of Business Economics and Management*, 10(4), 279–89.

Amaratunga, D., Baldry, D. and Sarshan, M. (2001) Process improvement through performance measurement: the balanced scorecard methodology. *Work Study*, 50(5), 179–88.

Ankrah, N. A. and Langford, D. A. (2005) Architects and contractors: a comparative study of organizational cultures. *Construction Management and Economics*, 23(6), 595–607.

Arbukle, J. L. and Wothke, W. (1999) *Amos 4.0 User's Guide*. Small Waters Corporation.

Atkinson, R., Crawford, L. and Ward, S. (2006) Fundamental uncertainties in projects and the scope of project management. *International Journal of Project Management*, 24, 687–98.

Barney, J. (1986) Strategic factor markets: expectations, luck, and business strategy. *Management Science*, 32(10), 1231–41.

Bayliss, R., Cheung, S. O., Suen, C. H. H. and Wong, S. P. (2004) Effective partnering tools in construction: A case study on MTRC TKE Contract 604 in Hong Kong. *International Journal of Project Management*, 22(3), 253–63.

Butler, J. K. (1991) Toward understanding and measuring conditions of trust: evolution of a conditions of trust inventory. *Journal of Management*, 17(3), 643–65.

Caerteling, J. S., Hartmann, A. and Tijhuis W. (2006) Innovation processes in construction: two case studies. *The Joint International Conference on Construction Culture, Innovation and Management*, Dubai, UAE. 26–29 November, 10–17.

Cameron, K. S. and Quinn, R. E. (1999) *Diagnosing and Changing Organizational Culture based on the Competing Values Framework*. Reading: Addison-Wesley.

Cheng, C. W. M. and Liu, A. M. M. (2007) The relationship of organizational culture and the implementation of total quality management in construction firms. *Surveying and Build Environment,* 18(1), 7–16.

Cheung, S. O., Wong, S. P. and Wu, W. Y. (2010) Towards an organizational culture framework in construction. *International Journal of Project Management,* 29(1), 33–44.

Cole, G. A. (1997) *Personnel Management* (4th edn). London: Letts Educational.

Construction Industry Review Committee of Hong Kong (CIRC) (2001) *Construct for Excellence.* Hong Kong: Construction Industry Review Committee.

Cummings, L. L. and Bromiley, P. (1996) The organizational trust inventory, in: R. Kramer and T. Tyler (eds), *Trust in Organisation, Frontier of Theory and Research.* Beverly Hills: Sage, 302–30.

Denison, D. R. (1990) *Corporate Culture and Organizational Effectiveness.* New York: Wiley.

Egan, J. (1998) *Rethinking Construction.* London: Department of the Environment, Transport and the Regions.

Gordon, G. G. and DiTomaso, N. (1992) Predicting corporate performance from organizational culture. *Journal of Management Studies,* 29, 783–98.

Hair, J. F., Anderson, R. E., Tatham, R. L. and Black, W. C. (1998) *Multivariate Data Analysis* (5th edn). Englewood Cliffs, NJ: Prentice Hall.

Hammer, M. (2004) Deep change: how operational innovation can transform your company. *Harvard Business Review,* 82(9), 85–93.

Hartmann, A. (2006a) The role of organizational culture in motivating innovative behaviour in construction firms. *Construction Innovation,* 6(3), 159–72.

Hartmann, A. (2006b) The context of innovation management in construction firms. *Construction Management and Economics,* 24(6), 567–78.

Hofstede, G. (2001) *Culture's Consequences: Comparing Values, Behaviors, Institutions, and Organizations across Nations.* London: Sage Publications.

Jöreskog, K. and Sörbom, D. (1996) *LISREL 8: Users Reference Guide.* Hillsdale, NJ: Lawrence Erlbaum Associates.

Kagioglou, M., Cooper, R. and Aouad, G. (2001) Performance management in construction: a conceptual framework. *Construction Management and Economics,* 19(1), 85–95.

Kaplan, R. S. and Norton, D. P. (1992) The balanced scorecard: measures that drive performance. *Harvard Business* Review, 70(1), 71–79.

Kululanga, G. K., Edum-Fotwe, F. T. and McCaffer, R. (2001) Measuring construction contractors' organizational learning. *Building Research and Information,* 29(1), 21–29.

Law, K. M. Y. and Chuah, K. B. (2004) Project-based action as learning approach in learning organization: the theory and framework. *Team Performance Management,* 10(7/8), 178–86.

Liu, A. M. (1999) Culture in the Hong Kong real-estate profession: a trait approach. *Habitat International,* 23(3), 413–25.

Marrewijk, A. (2007) Managing project culture: the case of Environ Megaproject. *International Journal of Project Management,* 25, 290–99.

Mohamed, S. (2003) Scorecard approach to benchmarking organizational safety culture in construction. *Journal of Construction Engineering and Management ASCE,* 129(1), 80–88.

Molenaar, K., Washington, S. and Diekmann, J. (2000) Structural equation model of construction contract dispute potential. *Journal of Construction Engineering and Management*, 126(4), 268–77.

Ozorhon, B., Arditi, D., Dikmen, I. and Birgonul, M. T. (2008) Implications of culture in the performance of international construction joint ventures. *Journal of Construction Engineering and Management*, 134(5), 361–70.

Parker, S. and Skitmore, M. (2005) Project management turnover: causes and effects on project performance. *International Journal of Project Management*, 23, 205–14.

Patanakul, P. and Milosevic, D. (2009) The effectiveness in managing a group of multiple projects: factors of influence and measurement criteria. *International Journal of Project Management*, 27, 216–33.

Peters, T. J. and Waterman, R. H. (1982) *In Search of Excellence: Lessons from America's Best-run Companies*. New York: Harper and Row.

Saffold, G. S. (1988) Culture traits, strength and organizational performance: moving beyond strong culture. *Academy of Management Review*, 13(4), 546–58.

Schein, E. H. (1985) *Organizational Culture and Leadership: A Dynamic View*. San Francisco: Jossey-Bass.

Schein, E. H. (1986) *Organizational Culture and Leadership*. San Francisco: Jossey-Bass.

Schein, E. H. (2004) *Organizational Culture and Leadership* (3rd edn). San Francisco: Jossey-Bass.

Sexton, M. and Barrett, P. (2003) Appropriate innovation in small construction firms. *Construction Management and Economics*, 21(6), 623–33.

Sharma, S. (1996) *Applied Multivariate Techniques*. New York: John Wiley & Sons.

Smircich, L. (1983) Concepts of culture and organizational analysis. *Administrative Science Quarterly*, 28, 339–58.

Winch, G. M. (2003) How innovative is construction? Comparing aggregated data on construction innovation and other sectors: a case of apples and pears. *Construction Management and Economics*, 21(6), 651–54.

Wong, P. S. P. and Cheung, S. O. (2005) Structural equation model of trust and partnering success. *Journal of Management in Engineering*, 21(2), 70–80.

Wong, P. S. P., Cheung, S. O. and Leung, K. Y. (2008) The moderating effect of organizational learning type on performance improvement. *Journal of Management in Engineering*, 24(3), 162–72.

Xiao, H. and Proverbs, D. (2003) Factors influencing contractor performance: an international investigation. *Engineering Construction and Architectural Management*, 10(5), 322–32.

Zeng, S. X., Xie, X. M., Tam, C. M. and Sun, P. M. (2009) Identifying cultural difference in R&D project for performance improvement: a field study. *Journal of Business Economics and Management*, 10(1), 61–70.

Zhang, S. B. and Liu, M. M. (2006) Organizational culture profiles of construction enterprises in China. *Construction Management and Economics*, 24(8), 817–28.

3 Engendering trust in construction contracting

Sai On Cheung, Wei Kei Wong and Hoi Yan Pang

Introduction

How trust establishes and sustains relationship has been widely studied in the fields of social science (Kramer 1999; Luhmann 1979; Lewis and Weigert 1985; Rousseau *et al.* 1998), economics (Das and Teng 2004; Zucker 1986) and business and management (Hartman 2000; McAllister 1995; McKnight *et al.* 1998; Whitener *et al.* 1998). It has been suggested that trust helps to reinforce individuals' affirmative willingness, confidence, expectation, belief and behaviour and to overcome risk/uncertainty (Caldwell and Clapham 2003; Kanawattanachai and Yoo 2002; McAllister 1995; Whitener *et al.* 1998; Zaghloul and Hartman 2003). All these are contributive in maintaining amicable relationship as a trustful environment would help to bridge gaps, establish faith and synergise the strengths of members of an organisation. According to Whitener *et al.* (1998), the presence of trust can also foster organisation solidity through 'fair' play. Trust prompts members to have faith in the organisation and buy in its policies and procedures to create a collegial working environment. Likewise, trust-building is one of the key competences to manage the relationship among contracting parties, particularly in the context of cooperative contracting (Bayliss *et al.* 2004; Black *et al.* 2000; Cheung *et al.* 2003; Cheung 2007; Cook and Hancher 1990; Hancher 1989; Wong and Cheung 2004; Wong and Cheung 2005; Wong *et al.* 2005). Although trust has been recognised as a relationship lubricant, its existence is often questioned. Operating these concepts for use in construction contracting shall be instrumental. This study aims to fill the gap and has three stages of work:

1 reviewing trust-related literatures;
2 proposing a trust framework in construction contracting;
3 testing of the proposed trust framework.

The study

Stage one: review on trust-related literatures

In construction, trust-building is often affiliated with the spirit of partnering. Hancher (1989) promoted the use of partnering instead of a traditional approach as a means to improve contracting relationships. Cook and Hancher (1990) built upon the work of Hancher and suggested that information sharing, such as exchanging organisational strategies or confidential information, is the essential trust-builder. It was reported by Cook and Hancher (1990) that appropriate and honest information sharing can optimise mutual understanding and expectations between the partnering members. In general, partnering projects achieve better project quality and safety, create new directions in technology usage and make more business. 'Mutual trust' has been found to be one of the most important success factors in maintaining partnering relationship (Black *et al.* 2000). In the partnering studies by Bayliss *et al.* (2004), Wong and Cheung (2004; 2005), Wong *et al.* (2005) and Cheung (2007), it was identified that the trust level between the client and contractor grows if trusting acts can be reciprocated. In their research with public sector organisations in Singapore, Wong *et al.* (2000) obtained empirical evidence to support that performance, acting with integrity and demonstrating concern are antecedent to trust. Zaghloul and Hartman (2003) proposed the inclusion of competence, integrity and intuition as trust measures in the construction industry. However, their proposition was based only on interpersonal trust and lacks objective solutions on risk allocation. Other trust-related studies in this field include those conducted by Huemer (2004) and Kadefors (2004). Huemer (2004) associated trust-predictability with the redefinition of roles and relationships in construction projects. Kadefors (2004) proposed economic incentives, traditional contractual arrangements and informal cooperative relationships as the factors that influence trust development in construction projects. Chan *et al.* (2010) provided valuable case studies on managing relationships in contracting. Previous research on the contribution of trust in construction contracting therefore appears to direct discussions in the improvements of time, cost and quality of construction projects, the relaxation of adversarial relationships and the promotion of cooperation. There is no shortage of theoretical deliberations on trust contributions, however, empirical evidence is uncommon. More significantly, little has been offered in conceptualising these trust models in the context of construction contracting. Stage Two of the study aims to achieve this research gap.

Stage two: proposing a trust framework in construction contracting

A trust framework in construction contracting is proposed by trusting behaviours manifesting the trust types identified in Stage One. Trusting behaviours involve elements of expectation, confidence, willingness, belief, behaviour (McAllister 1995; Whitener *et al.* 1998), reliance, hopefulness, optimism, honesty, mutuality, dependency, sharing of values, reciprocity, commitment, caring, responsibility (Das and Teng 2004; Lewis and Weigert 1985; McAllister 1995; Whitener *et al.* 1998), uncertainty and risk (Das and Teng 2004). In the light of the literature review on trust, it is proposed that trust be categorised into system-based, cognition-based and affect-based. Table 3.1 gives a summary of the trust types expounded in previous studies.

Table 3.1 Classification of trust

Authors	Definitions and scopes		Dimension of trust
Luhmann (1979)	Personal trust	Personal trust involves an emotional bond between individuals. Its emotional component acts as a protective base of trust when experiencing betrayal or destructive events.	Cognition-based and Affect-based
	System trust	System trust contains no emotional content. It rests on a presentational base and it is essential for the effective function of money or power exchange.	System-based
Lewis and Weigert (1985)	System trust	System trust refers to trust in the functioning of bureaucratic sanctions and safeguards in terms of the legal system.	System-based
	Cognitive trust	This trust is a combination of low emotionality and high rationality and is based on some 'good reasons' constituting evidence of trustworthiness.	Cognition-based
	Emotional trust	A mix of high emotionality and low rationality constitutes emotional trust which involves bonds of friendship and love.	Affect-based
McAllister (1995)	Affect-based trust	Affect-based trust refers to the intention to provide extra help and assistance that is outside an individual's work role without remuneration.	Affect-based
	Cognition-based trust	Cognition-based trust develops based on the success of past interaction, the extent of similarity, and organisational context considerations.	Cognition-based

Table 3.1 continued

Authors	Definitions and scopes		Dimension of trust
Rousseau *et al.* (1998)	Calculus-based trust	This trust describes the perceptions about one to another regarding some beneficial issues. References, certificates and diplomas are the media promoting calculus-based trust.	Cognition-based
	Relational trust	This trust is arisen by continual interactions between individuals. Emotions and personal attachments are also influential to the trusting relationship.	Affect-based
	Institution-based trust	Legal systems, conflict management and co-operation, systems regulating education and professional practice were suggested as tools to sharp trust in institutions.	System-based
Hartman (2000)	Competence trust	Competence trust is defined to be based on one's perception of the other's capacity to perform.	Cognition-based
	Integrity trust	Integrity trust is found upon one's perception of the other's attitude to act ethically and be motivated not to take one's advantage.	Cognition-based
	Intuitive trust	Intuitive trust involves emotion and intuition about one's impression to the other.	Affect-based
Kramer (1999)	Dispositional trust	This trust develops based on the build-up of general belief on early trust-related experience.	Affect-based
	History-based trust	Individuals develop history-based trust based on previous interactional information and experience.	Cognition-based
	Third parties as conduits of trust	Individuals adopt second-hand knowledge in order to assess the other's trustworthiness.	Cognition-based
	Category-based trust	This trust develops according to knowledge acquired from one's membership in a social or organisational category.	Cognition-based
	Role-based trust	Role-based trust grows based upon knowledge of role relations, rather than specific knowledge about one's capabilities, dispositions, motives and intentions.	Cognition-based
	Rule-based trust	Rule-based trust is subject to shared understandings of the system of rules concerning appropriate behaviour.	System-based

System-based trust

System-based trust focuses on formalised and procedural arrangements with no consideration of personal issues (Lewis and Weigert 1985). These arrangements regulate organisational behaviours and engender trust through uncertainty reduction (Whitener *et al.* 1998). To develop system-based trust, organisational policy, communication systems and contracts/agreements are the three major attributes.

ORGANISATIONAL POLICY

Organisational policy specifies priorities and explains business procedures. It reflects an organisation's value system. Organisational policy reflects the expected behaviour of the staff and the trust they have in the organisation. Nonetheless, McKnight *et al.* (1998) suggested that system-based trust can be developed and strengthened through one's belief in the organisational policy in achieving organisational goals. Sufficiently flexible systems embracing innovative management, congruent organisational structures and processes together with appropriate organisational culture provide the platform for trust development and maintenance (Zaghloul and Hartman 2003; Kadefors 2004). Organisational policy is therefore a key nutrient for the development of system-based trust.

COMMUNICATION SYSTEM

Communication system defines the channels for interactions within an organisation. Such interactions can be identified as either close or distant contacts. Meetings, workshops or visits are examples of close contacts while emails, telephones and teleconferencing typify distant contacts. According to Gayeski (1993), system-based trust can be enhanced through a clearly defined system of communication procedures and approaches. Not only does it facilitate convenient and speedy communications, it also reduces arguments that often stem from misunderstandings resulting from untimely and inaccurate information exchanges. As suggested by Zaghloul and Hartman (2003) and Wong and Cheung (2004), a good communication system mitigates risks and increases reputation of all concerned parties. An efficient communication system is therefore another indicator of system-based trust.

CONTRACTS/AGREEMENTS

Contracts and agreements define relationships and obligations between individuals (McKnight *et al.* 1998; Zaghloul and Hartman 2003) and are regarded as another attribute of system-based trust because of their ability to reduce uncertainties and minimise, share or shift risks between contracting parties (Das and Teng 2004). Contracts and agreements explicate implicit expectations and make obligations and rights visible. This contributes to fair risk allocation, overall project performance improvement (Kadefors 2004) and cost reduction (Kramer 1999). With these, building system-based trust becomes possible.

Cognition-based trust

Cognition-based trust develops from the confidence built upon knowledge that reveals the cognitive bearings of an individual or an organisation. The exchange of such knowledge can be attained by formal and informal communicative exchanges.

COMMUNICATION/INTERACTION

Communication/interaction is a means of imparting or exchanging information between individuals or organisations (Zucker 1986). Continuing communication and interaction facilitate organisational members to distribute, comprehend and obtain information that can be translated into meaningful knowledge (Gayeski 1993). The absence of communication creates fear of exploitation and betrayal, which would result in avoidance of commitment to the team. Communication/interaction fosters mutual trust, which lays the foundation for growth in trust and business relationship (Das and Teng 2004; McAllister 1995). Cognition-based trust brings about collaborative efforts that underpin project success. It also enables the working members to learn about the needs and capabilities of their partners. In the construction industry, communication has been identified as an effective means to reduce conflict (Bayliss *et al.* 2004).

KNOWLEDGE

Knowledge can be translated from information. Knowledge, such as track record, organisational role and financial status, reveals the consistency, competence as well as integrity of the individual or organisation. This type of knowledge is critical in cognition-based trust development (Caldwell and Clapham 2003; Huemer 2004; Kanawattanachai and Yoo 2002; Wong and Cheung 2004). For example, Kramer (1999) advocated that track record enables cognition-based trust-building resulting from information acquisition and relationship enhancement. A person's role in an organisation, his track record, reputation and professional standing are essential indicators of one's trustworthiness (Kramer 1999; McAllister 1995). Having the knowledge of past performance and reputation assures the relevancy and accuracy of the assessment of the reliability of this person (Zucker 1986). Financial status comprises promises of the company's commitment in promoting cognition-based trust (Kramer 1999) because it reflects the company's capability to facilitate and enhance economic processes, manage risks and absorb shocks. In the construction industry, knowledge as discussed is essential for cognition-based trust building and toward project success.

Affect-based trust

Affect-based trust is built on a sentimental platform. It describes an emotional bond that ties individuals to invest in personal attachment and be thoughtful to each other (Lewis and Weigert 1985). Although Boon and Holmes (1991) suggested affect-based trust development is restricted to romantic relationships, it is believed that building up affect-based trust at work enhances the process of evaluation and information exchange, improves performance and wellbeing of the teams (McAllister 1995). Therefore, being thoughtful and emotional investments are proposed to describe affect-based trust development.

BEING THOUGHTFUL

Being thoughtful can be demonstrated by showing care and concern (McAllister 1995). It eliminates unfavourable attitudes and raises kind awareness of other people's feelings (Kanawattanachai and Yoo 2002; Wright 1996). Wright (1996) noted the reciprocal nature of thoughtfulness which makes it an effective ingredient for improving work relationships and developing affect-based trust.

EMOTIONAL INVESTMENTS

Emotion is an affective state of consciousness often actuated by personal feelings. It is a mental state that happens spontaneously instead of through cognitive or volitional effort and is often accompanied by physiological changes. The attempt of an individual to make emotional investment illustrates his enthusiasm on spending time, energy and effort on a person that he thinks is good or helpful. Therefore, an individual's willingness to invest his emotions on others demonstrates affect-based trust building. Affect-based trust derived from emotional investments reduces defensiveness, unhealthy competitiveness and disruption, eliminates frictions and enhances team spirit and morale in working relationship (Boon and Holmes 1991).

In the light of the above, a framework for trust in construction contracting is proposed and presented in Figure 3.1. The three forms of trust and their attributes are included in this framework. Stage Three of the study aims to confirm the underlying constructs of the proposed framework.

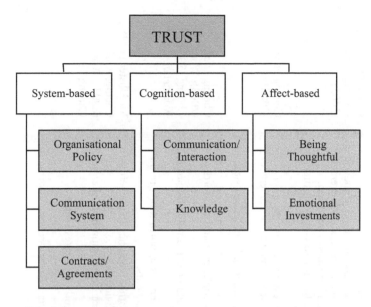

Figure 3.1 A framework for trust in construction contracting

Stage three: testing of the proposed trust framework

This stage of the study aims to test the underlying constructs of the proposed trust framework. First, data was collected through a questionnaire survey with construction practitioners. With these, a structural equation modelling analysis was then conducted.

Data collection

The trust attitude in construction contracting was collected through a questionnaire survey. The questionnaire was designed based on the proposed trust framework as described in Stages One and Two of the study. Twenty-three trusting behaviours were long-listed. These are arranged trust types and their characterising behaviours as presented in Table 3.2. This stage of the study is important in establishing the empirical bases, thus enhancing the authenticity of the framework. The analytical results will also provide insight in the roles played by various forms of trust.

The respondents were asked to assess the degree of agreement on a seven-point Likert scale against each of the twenty-three statements. A higher score represents a higher level of agreement in the statement. In this study, a total of 467 questionnaires were sent to project managers, architects, engineers, contract/legal advisers, quantity surveyors and project coordinators.

Table 3.2 Trust types and their characterising behaviours

Trust types	Attributes	Trusting behaviours	References
System-based	Organisational policy	II8. Sufficient organisational resources in response to contracting parties' needs increase the sense of belonging to the organisation. II12. An organisation should clearly define the job tasks required of individuals. II14. Good management of organisation system reinforces goal achievement such as continual improvement, profit making and business expansion. II23. Organisation policy should be clearly specified for solving cost, time, risk and safety issues.	Caldwell and Clapham (2003); Cheung et al. (2003).
	Communication system	II4. Using effective communication methods are essential at work. II9. Formal communications with working partners should be documented in a systematic way. II21. Keeping a good communication system would avoid ambiguous situations and discrepancies occurring at all times.	Cheung et al. (2003).
	Contracts and agreements	II2. A clearly defined contract document brings confidence and comforts to all parties. II15. Information in the contract document should be explainable to parties who may be affected. II16. Clarification of contract terms and agreements is important before the commencement of work to minimise future arguments.	Cheung et al. (2003); Wong and Cheung (2004); Wong et al. (2005).
Cognition-based	Communication/ interaction	II1. Keeping a long-term relationship with the other party has the benefit of maintaining better communication between individuals. II10. Good interaction allows me to obtain more information from the other party. II13. Attending work-related interaction frequently facilitates better understanding between individuals. II18. Open and honest communication enables more work-related information exchange between individuals.	McAllister (1995); Cheung et al. (2003); Wong and Cheung (2004).

Table 3.2 continued

	Knowledge	II3. Track record is an essential tool to judge the other party's competence and consistency level. II5. Financial stability is one of the factors in evaluating a company's reliability. II7. The other party will have confidence to work with me if I have a good reputation of being honest.	McAllister (1995); Wong and Cheung (2004).
Affect-based	Being thoughtful	II9. Showing care and concern to my workmate at appropriate time impresses his/her feeling more comfortable to work with me. II20. Being considerate is a tool to understand an individual's needs and feeling at work so as to achieve his/her maximum capacity. II22. Taking each party's needs into account in decision-making process encourages a compromising and satisfactory outcome.	McAllister (1995); Kanawattanachai and Yoo (2002).
	Emotional investments	II6. Having a good personal relationship with the other party may also improve working relationship with him/her. II11. I am more likely to rely on a working partner whom I have good impression. II17. Spending appropriate time, energy and effort to understand other party's personal details and work background eliminates frictions between each other at work.	McAllister (1995); Zaghloul and Hartman (2003).

They were identified from the government and professional directories and websites of companies. 163 of them responded to the survey, which represented a response rate of 34.9 per cent. The respondents are experienced practitioners with over 70 per cent of them having more than ten-years experience in this field. The study sample represented a broad spectrum of professional disciplines working as client, consultant or contractor. These profiles are shown in Tables 3.3 and 3.4.

The framework constructs

The construct of the trust framework is tested by a structural equation model (SEM). This technique seeks to determine whether the number of factors and the loadings of measured variables on them conform to what is postulated in theory (Hair *et al*. 1998). With this method, each equation in the SEM model represents a causal link rather than an empirical association (Jöreskog 1982). Furthermore, Goldberger (1973) presented three situations to demonstrate the advantages of SEM over the traditional multiple regression model:

> (1) when the observed measurements contain measurement errors and when the interesting relationship is among the true or dis-attenuated variables; (2) when there is interdependence or simultaneous causation among the observed response variables and (3) when important explanatory variables have not been observed.

Table 3.3 Profile of respondents (by organisational type)

Organisational types	Percentage
Client	29.45
Consultant	27.61
Contractor	42.94
	100.00

Table 3.4 Profile of respondents (by profession)

Professions	Percentage
Project manager	17.79
Architect	7.97
Engineer	39.88
Contract/legal adviser	3.07
Quantity surveyor	26.38
Project coordinator	4.91
	100.00

In these regards, SEM is used for this study instead multiple regression. In an SEM model, appropriate goodness-of-fit indices of structural equation modelling are used to confirm the 'fitness' of the framework. Model fit indicators include relative chi-square ($\chi2/df$) < 2.00 (Bollen 1989), goodness of fit index (GFI) > 0.80 (Maskarinec *et al.* 2000), comparative fit index (CFI) > 0.80 (Maskarinec *et al.* 2000), Tucker-Lewis coefficient (TLI) > 0.80 (Maskarinec *et al.* 2000) and root mean square error of approximation (RMSEA) < 0.08 (Hair *et al.* 1998). Moreover, the relationship of each variable can be determined by the use of path analysis that estimates the strength of each relationship in a model (Hair *et al.* 1998). A path coefficient is a standardised regression weight to be considered when discussing the regression. It shows the direct effect between a pair of independent variables and dependent variables in the path model. The greater the path coefficients, the stronger the evidence that the measured variables are representing the underlying paradigms (Bollen 1989). The advantage of using path coefficients is simplicity of exposition.

STATISTICAL FIT OF THE FRAMEWORK

In order to perform the confirmatory factor analysis, the trust framework was arranged as a structural equation model as shown in Figure 3.2. The five goodness-of-fit indices were obtained by running Amos 5, a software program designed for conducting structural equation modelling. The goodness-of-fit indices obtained are $\chi2/df$ = 1.89, GFI = 0.81, CFI = 0.89, TLI = 0.88 and RMSEA = 0.07. Comparing to the statistical requirements of goodness-of-fit indices, the results were considered satisfactory. Therefore, the proposed trust framework is supported.

Path analysis was then carried out to determine the relationship among the variables. Figure 3.3 shows the output of the path analysis, which shows that the path coefficients for system-based, cognition-based and affect-based trust are 0.97, 0.99 and 0.94 respectively. Comparing the path coefficients of these three forms, cognition-based trust scored the highest, which suggests cognition-based trust to be the most important form of trust among the three. The two attributes of cognition-based trust also have high path coefficients. Path coefficient between communication/interaction and cognition-based trust was 0.99 and that between knowledge and cognition-based trust was 0.95.

Discussion

Cognition-based trust describes a trusting relationship that builds on mutual understanding through fruitful information exchange and acquaintance. Construction projects require team members of different expertise working together. Thus, good cooperation is desirable in order to complete a project at minimum cost, least time and best quality. Communication/

interaction forms the bridge for daily information exchanges because working members have to rely on what they have been provided with. Such reliance enables trust building. It is therefore not surprising that keeping good communication/interaction develops cognition-based trust. Moreover, information is also very important in the construction industry, especially at the planning stage of construction projects. Clients place much attention on and acquire information from the record of consultants and contractors. In the construction industry in Hong Kong, the government has a system of consultants' performance evaluation and selection of contractors. The consultants' past performance is reported and evaluated. Unsatisfactory performance will lead to suspension from bidding for future consultancies. The tender evaluation of contractors is reduced to scores that will be given according to the tenderer's experience, past performance, technical resources and technical proposal. These assessments collectively account for 40 per cent of the total assessment, and hence can be decisive in bid evaluation. It can be seen that construction practitioners have great interest in getting hold of information on each other because of the benchmarking function. Consequently, a company or an individual who has a wealth of information that can be translated into knowledge is more likely to be trusted by others.

System-based trust has the second highest path coefficient. Its attributes – organisational policy, communication system and contracts/agreements – also have high path coefficients with system-based trust. In construction, a contract document that includes conditions of contract, specifications, bills of quantities and contract drawings details the rights and obligations of the contracting parties. This is also instrumental in facilitating system-based trust. Basically, achieving the stipulated requirement is the first and foremost step in deriving system-based trust; this is because the systems are stipulated in the contract, in particular those related to performance. Fulfilment of the system requirements brings about the confidence of the contracting parties and therefore activates trust. Effective communication is pivotal to the smooth running of a project. Regular contacts, such as progress meetings, technical meetings, quality control workshops, site safety workshops or site visits are typically scheduled so that performance can be monitored. Other channels, such as emails and teleconferencing, have become more and more important and efficient means for information exchange.

Organisational policy describes the strategic settings of an organisation, and thus guides decision making in business execution. In other words, an organisation speaks and acts through its policies. Thus, policy has a major impact on how organisational members view the company. This perception influences organisational members' sense of identification with the company and its objectives. An enduring organisational policy is important to any organisation as it can be viewed as its identity. Construction work often requires group efforts whereby proper administration guidelines help to connect team members and develop trust among them. Organisational policy therefore becomes a critical system of the working members' performance guidebook.

Affect-based trust is comparatively the least influential among the three forms of trust (path coefficient 0.94). Nonetheless, its attributes of being thoughtful and emotional investments have high path coefficients of 0.92 and 0.99 respectively. Both of these attributes are grounded on intuitive elements and relate to personal feelings and subjective decisions. However, construction professionals should control personal feelings so that they can exercise reasonable skills and care and perform up to standard. Therefore, investment of emotion should be minimised at work. The suggestion does not negate the existence of affect-based trust in the construction industry, as showing certain levels of care, concern and being considerate would help in promoting a trusting working relationship (Wright 1996).

Summary

Trust has been identified as the most important behavioural factor in managing relationship. In construction, where collaboration among contracting parties is essential in order to accomplish sophisticated tasks that require multi-parties' involvement, successful trust-building within project teams would certainly improve the project outcome. Despite this obvious advantage, trust appears to be a stranger in construction contracting where confrontation remains the prevalent environment.

This study aims to further the understanding of trust in construction contracting. First, trust concepts were reviewed. Categorisation was then developed to put these conceptions in perspective. A trust framework in construction contracting was then developed by reducing these trust conceptions to behavioural statements. The constructs of the framework was then tested empirically through the technique of structural equation modelling. The study identified three broad types of trust: system-based, cognition-based and affect-based. Cognition-based trust is built on knowledge and understanding. System-based trust is founded on performance and faith in the system. Affect-based trust appears to address the feelings and emotions, and thus tends to be more personal. The empirical results suggested that all three forms are of almost equal importance in trust building. Trust building is more easy to say than do. The three facets of trust co-exist and in fact are mutually dependent. A system is only as good as its weakest point, hence an organisation must be able to install robust systems and care for its members. The trust framework thus enhances our outstanding in how trust building can be practiced in construction contracting.

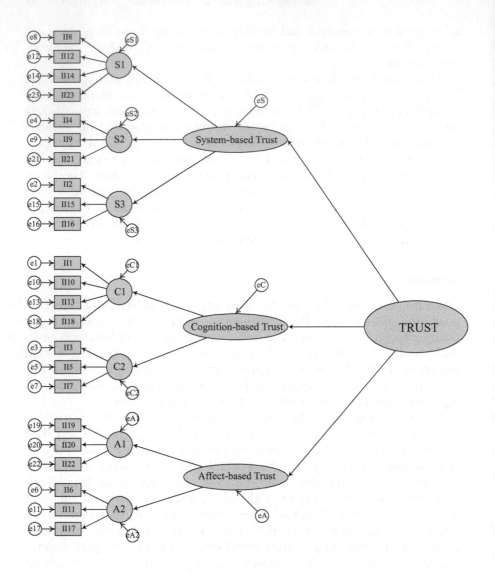

Figure 3.2 A structural equation model of the trust framework in construction contracting

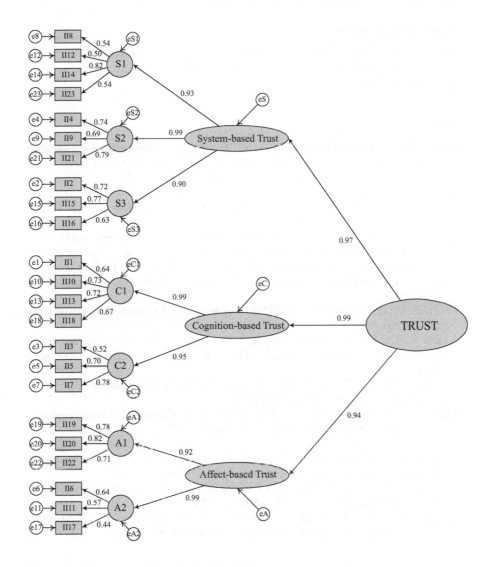

Figure 3.3 Output of the path analysis

Acknowledgements

The content of this chapter has been published in Volume 26(8) of *the International Journal of Project Management* and is used with the permission from Elsevier.

References

Bayliss, R., Cheung, S. O., Suen, C. H. and Wong, S. P. (2004) Effective partnering tools in construction: a case study on MTRC TKE Contract 604 in Hong Kong. *International Journal of Project Management*, 22(3), 253–63.

Black, C., Akintoye, A. and Fitzgerald, E. (2000) An analysis of success factors and benefits of partnering in construction. *International Journal of Project Management*, 18(6), 423–34.

Bollen, K. A. (1989) *Structural Equations with Latent Variables*. New York: Wiley.

Boon, S. D. and Holmes, J. G. (1991) The dynamics of interpersonal trust: resolving uncertainty in the face of risk, in R. A. Hinde and J. Groebel (eds), *Cooperation and Prosocial Behavior*. Cambridge: Cambridge University Press, 190–211.

Caldwell, C. and Clapham, S. E. (2003) Organizational trustworthiness: an international perspective. *Journal of Business Ethics*, 47(4), 349–64.

Chan, A. P., Chan, D. W. and Yeung, J. F. (2010) *Relational Contracting for Construction Excellence: Principles, Practices and Case Studies*. London: Spon Press.

Cheung, S. O. (2007) *Trust in Co-operative Contracting in Construction*. Hong Kong: City University of Hong Kong Press.

Cheung, S. O., Ng, S. T., Wong, S. P. and Suen, C. H. (2003) Behavioral Aspects in Construction Partnering. *International Journal of Project Management*, 21(5), 333–43.

Cook, E. L. and Hancher, D. E. (1990) Partnering: contracting for the future. *Journal of Management in Engineering*, 6(4), 431–46.

Das, T. K. and Teng, B. S. (2004) The Risk-based view of trust: a conceptual framework. *Journal of Business and Psychology*, 19(1), 85–116.

Gayeski, D. (1993) *Corporate Communications Management: The Renaissance Communicator in Information-Age Organizations*. Boston, MA: Focal Press/Heinemann.

Goldberger, A. S. (1973) Structural equation models: an overview, in A. S. Goldberger and O. D. Duncan (eds), *Structural Equation Models in the Social Sciences*. New York: Seminar Press, 1–18.

Hair, J. F., Anderson, R. E., Tatham, R. L. and Black, W. C. (1998) *Multivariate Data Analysis*. Englewood Cliffs, NJ: Prentice Hall.

Hancher, D. E. (1989) Partnering: meeting the challenges of the future. *Interim Report of the Task Force on Partnering*, Construction Industry Institute.

Hartman, F. (2000) The role of trust in project management, in *Proceedings of the PMI Research Conference*, Project Management Institute, Pennsylvania, USA.

Huemer, L. (2004) Activating trust: the redefinition of roles and relationships in an international construction project. *International Marketing Review*, 21(2), 187–201.

Jöreskog, K. (1982) Recent developments in structural equation modelling. *Journal of Marketing Research*, 19(4), 404–16.

Kadefors, A. (2004) Trust in project relationships: inside the black box. *International Journal of Project Management*, 22(3), 175–82.

Kanawattanachai, P. and Yoo, Y. (2002) Dynamic nature of trust in virtual teams. *Journal of Strategic Information Systems*, 11(3–4), 187–213.

Kramer, R. M. (1999) Trust and distrust in organizations: emerging perspectives, enduring questions. *Annual Review of Psychology*, 50, 569–98.

Lewis, J. D., Weigert, A. (1985) Trust as a social reality. *Social Forces*, 63(4), 967–85.

Luhmann, N. (1979) *Trust and Power*. New York: Wiley.

Maskarinec, G., Novotny, R. and Tasaki, K. (2000) Dietary patterns are associated with Body Mass Index in multiethnic women. *Journal of Nutrition*, 130, 3068–72.

McAllister, D. J. (1995) Affect- and cognition-based trust as foundations for inter-personal cooperation in organizations. *Academy of Management Journal*, 38(1), 24–59.

McKnight, D. H., Cummings, L. L. and Chervany, N. L. (1998) Initial trust formation in new organizational relationships. *The Academy of Management Review*, 23(3), 473–90.

Rousseau, D. M., Sitkin, S. B., Burt, R. S. and Camerer, C. (1998) Not so different after all: a cross-discipline view of trust. *Academy of Management Review*, 23(3), 393–404.

Whitener, E. M., Brodt, S. E., Korsgaard, M. A. and Werner, J. M. (1998) Managers as initiators of trust: an exchange relationship framework for understanding managerial trustworthy behavior. *The Academy of Management Review*, 23(3), 513–30.

Wong, E. S., Then, D. and Skitmore, M. (2000) Antecedents of trust in intra-organizational relationships within three Singapore public sector construction project management agencies. *Construction Management and Economics*, 18(7), 797–806.

Wong, S. P. and Cheung, S. O. (2004) Trust in construction partnering: views from parties of the partnering dance. *International Journal of Project Management*, 22(6), 437–46.

Wong, S. P. and Cheung, S. O. (2005) Structural equation model on trust and partnering success. *Journal of Management in Engineering*, 21(2), 70–80.

Wong, S. P., Cheung, S. O. and Ho, K. M. (2005) Contractor as trust initiator in construction partnering: a prisoner's dilemma perspective. *Journal of Construction Engineering and Management*, ASCE, 131(10), 1045–53.

Wright, A. D. (1996) Cross-cultural exploration of attitudes toward leader-subordinate friendship, Thesis for D.B.A., The University of Memphis.

Zaghloul, R. and Hartman, F. (2003) Construction contracts: the cost of mistrust. *International Journal of Project Management*, 21(6), 419–24.

Zuckerk, L. G. (1986) The production of trust: institutional sources of economic structure, in B. M. Staw amd L. l. Cummings (eds), *Research Organizational Behavior*. Greenwich, CT: JAI Press, 55–111.

4 The aggressive-cooperative drivers of construction contracting

Sai On Cheung and Tak Wing Yiu

Introduction

Construction contracting behaviour (CCB) is the attitude taken by contracting parties in performing a construction contract. It reflects the contracting parties' attitude and expectations in construction contracting transactions. To this end, cooperative contracting behaviour has long been promoted in view of the perceived benefits. This is because a cooperative working environment can maintain a harmonious relationship among contracting parties, and can allow effective enforcement of contractual rights and obligations (Harmon 2003; Yiu and Cheung 2006). However, the reality is that conflicts are inherent in most construction projects (Bramble and Cipollini 1995; Fenn *et al.* 1997; Pinnell 1999; Yiu and Cheung 2006; Zack 1995). Construction contracting behaviour remains largely adversarial as reported in a number of industrial reviews (CIRC 2001; Egan 1998; Latham 1994). In this connection, a stream of studies conducted by the construction community has also affirmed the need to overhaul the adversarial approach. These studies include case studies (Bayliss *et al.* 2004; Bennett and Jayes 1995; Black *et al.* 1999; Cheung and Suen 2002; Cheung *et al.* 2003; Sanvido *et al.* 1992) and identification of critical success factors (AGCA 1991; CIIA 1996; DeVilbiss 2000; Kumaraswamy and Matthews 2000; Li *et al.* 2000). It has generally been found that adversarial behaviour undermines cooperation among contracting parties and goes against amicable completion of construction projects (Byrnes 2002; Harmon 2001; Harmon 2003). These studies also suggest that cooperation enables synergistic efforts to maximise common interest. It seems there is a mismatch between the confrontational practice and the preferred state of cooperation. This apparent divergence between practice and preference suggests the existence of drivers for both approaches. Drivers mean those strengths and stimuli that motivate cooperative or aggressive moves. This means contracting parties can take a cooperative or aggressive stance in pursuing their goals depending on the significance of the influence of the drivers. For example, where construction contracts are not fulfilling the intended role of establishing the contractual responsibilities of the contracting parties (Dozzi *et al.* 1996), opportunistic aggressive moves may be adopted. Identification of the cooperative and aggressive drivers in construction contracting would help management to identify ways to prevent contracting parties from adopting aggressive moves during their contract administration or to facilitate an environment in fostering cooperative contracting. This is considered as one way of enhancing the soft power of the organisation. With these aims, a three-stage research work was designed in this study. Stage 1, entitled 'Identification of aggressive and cooperative drivers', aims to long-list the generic types of aggressive/cooperative drivers and their respective effects on CCB. Taxonomies of aggressive and cooperative drivers were developed in Stage 2 of the study. Based on the results obtained from Stages 1 and 2, Stage 3 involves the use of relative importance rankings to compare the significances of aggressive and cooperative drivers on CCB. Each of these stages is described seriatim.

Stage 1: Identification of aggressive and cooperative drivers of construction contracting behaviour

The aim of this Stage is to identify the generic types of aggressive and cooperative drivers of CCB. To achieve this, a questionnaire survey was employed to collect case-specific data. The questionnaire is divided into four main sections. The first section required the respondents to provide their background information and the particulars of their most recently completed construction project. The second section was designed to assess the respondents' degree of aggressiveness/cooperativeness in construction contracting behaviours. The respondents were asked to select the description that best reflects their contracting behaviour for their completed construction projects (Figure 4.1).

The self-reported contracting behaviour is then used in analysing the importance of aggressive and cooperative drivers in the final stage of this study. Further details are provided in the discussion section of the chapter. The next two sections were designed to address the aggressive and cooperative drivers of construction contracting behaviours. Comprehensive literature reviews were conducted to identify a list of aggressive drivers. A total of 17 (from AF-1 to AF-17) aggressive drivers were identified. The list of aggressive drivers is shown in Table 4.1. To facilitate interpretation, these drivers can be generally classified into five key attributes according to their

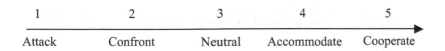

1	2	3	4	5
Attack	Confront	Neutral	Accommodate	Cooperate

Figure 4.1 The measurement scale of degree of aggressiveness/cooperativeness

Table 4.1 List of aggressive drivers

Coding	Attributes	Aggressive drivers	Reference
AF-1	Quality of previous dealing(s)	The quality of past dealing(s) between project participants was poor (low degree of satisfaction of previous cooperation)	Larson 1992; Parkhe 1993; Tallman and Shenkar 1994
AF-2		The previous dealing(s) were unsuccessful in achieving the goals of the project(s).	
AF-3	Level of competitive pressure	There had been great changes in the project content, due to environmental issues arising during the course of construction.	Amit *et al.* 1988; Hannan and Freeman 1984; Tushman and Romanelli 1985
AF-4		The actions being taken by competitors/ other contracting parties were strongly aggressive.	
AF-5		Dealing with the issues that can increase profitability would increase the competitive pressure of your project team.	
AF-6		The capital necessary for the project operation had been in general insufficient.	
AF-7	Intensity of competitive inertia	Your project team would become more active to deal with the issues that can benefit to achieve your goal.	Dutton and Duncan 1987; Mamer and McCardle 1987; Monahan 1987; Swedberg 1987; Macmillan and Jennifer 1990
AF-8		Having fair expectations of future profits and rewards would make your project team more likely to gain an advantage over the other parties.	
AF-9		Perception to aggressive actions of competitor/other contracting parties would more likely make your project team oppose.	
AF-10		The contract conditions were onerous and the performance specifications were harsh.	
AF-11		Complicated project had made your project team gain an advantage over the other parties.	
AF-12		Low interdependency between the project participants had led to your party more likely taking advantage over the others.	
AF-13	Likelihood of dispute(s)	The existence of exculpatory clauses had caused disputes in this project.	Goldberg 1992; Hartman 1993; Rubin and Brown 1975; Tallman and Shenkar 1994
AF-14	Likelihood of dispute(s), contract incomplete-ness	Not clearly stipulated contract conditions regarding the respective rights, benefits and responsibilities had caused disputes.	Goldberg 1992; Hackett 1993; Luo 2002

Table 4.1 continued

Coding	Attributes	Aggressive drivers	Reference
AF-15	Likelihood of dispute(s)	The unfavourable past cooperation between project participants had caused disputes.	Goldberg 1992; Hartman 1993; Rubin and
AF-16		Overly detailed contractual procedures to deal with contingencies had caused disputes.	Brown 1975; Tallman and Shenkar 1994
AF-17	Contract incompleteness	There were many ambiguous terms in the Conditions of Contract used.	Goldberg 1992; Hackett 1993; Luo 2002

natures: (1) quality of previous dealing(s); (2) level of competitive pressure; (3) intensity of competitive inertia; (4) likelihood of dispute(s); and (5) contract incompleteness. Furthermore, 27 cooperative drivers were extracted from the study of Yiu (2007). These were generally classified into 13 attributes: (1) teamwork intensity; (2) trust intensity; (3) effectiveness of communication; (4) goodness in relationship among contracting parties; (5) commitment maintenance; (6) goal mutuality; (7) availability of information; (8) involvement intensity; (9) incentive to risk savings or sharing; (10) effectiveness in dispute resolution; (11) effectiveness in solving/sharing of problems; (12) contract completeness; and (13) inter-party reciprocity. The list of cooperative drivers is shown in Table 4.2.

The drivers listed in Tables 4.1 and 4.2 were then used in the data collection of the questionnaire. Targeted respondents were identified from the Hong Kong Builder Directory and the websites of professional institutes such as the Hong Kong Institute of Surveyors (HKIS) and the Hong Kong Institution of Engineers (HKIE). A total of 300 questionnaires was sent, 100 of them were completed and returned, which represents a 33 per cent response rate. The respondents were construction professionals, including project managers (15 per cent), architects (15 per cent), engineers (25 per cent), quantity surveyors (42 per cent) and construction lawyers/mediators (3 per cent) from the government, consultancy firms and contractors. Over 57 per cent of them had at least 10 years' experience in construction.

Stage 2: taxonomies of aggressive and cooperative drivers

Based on the data collected from the questionnaire survey, this stage aims at developing taxonomies of aggressive and cooperative drivers of CCB. Taxonomy is a system by which categories are related to one another by means of class inclusion (Rosch 1988). The taxonomies of aggressive and

Table 4.2 List of cooperative drivers

Coding	Attributes	Cooperative drivers (Yiu 2007)
CF-1	Teamwork intensity	Teamwork spirit facilitated dispute resolution effectively.
CF-2	Trust intensity	Previous dealing(s) among project team members reinforced confidence in working with each other.
CF-3	Effectiveness of communication	Effective communication from previous dealing(s) among project team members.
CF-4	Teamwork intensity	Teamwork spirit among project team members facilitated project progress effectively.
CF-5	Goodness in relationship among	A good working relationship from previous dealing(s) among project team members.
CF-6	contracting parties	A good personal relationship among project team members.
CF-7	Teamwork intensity	Trust developed among project team members facilitated project progress effectively.
CF-8		Project team members were willing to share thoughts and feelings with each other.
CF-9		Project team member had open and honest communications.
CF-10	Commitment maintenance	A high degree of involvement within project team members.
CF-11		Your project team demonstrated open commitment within project team members.
CF-12	Goal mutuality	The contract design encouraged project team members to achieve common objectives in a rational manner.
CF-13	Availability of information	There were no constraints in getting information from other project team members.
CF-14	Goal mutuality	The project team members had mutual goals.
CF-15	Availability of information	Project team members had plenty of experience in handling project(s) of a similar nature.
CF-16		Good information exchange from previous dealing(s) among project team members.
CF-17	Involvement intensity	The involvement of your project team members had been highly voluntary.
CF-18	Incentive to risk savings or sharing	The contract provisions apportioned risks equitably between contracting parties.
CF-19		The contract provisions and specifications provided incentive for risk sharing among the contracting parties.
CF-20		Provision of tangible reward (e.g. bonus, a gain share fund, etc.) enabled the project team members to generate more incentive to save cost.
CF-21		A modest contract sum would make project team members less risk averse.
CF-22	Effectiveness in dispute resolution	Guidelines of handling various unanticipated contingencies had been incorporated in the contract in most cases.

Table 4.2 continued

Coding	Attributes	Cooperative drivers (Yiu 2007)
CF-23		Provision of a third party who was paid jointly by the contracting parties (e.g. an experienced mediator) enabled effective dispute resolution.
CF-24	Effectiveness in solving/sharing of problems	Mutual consultation among contracting parties enabled effective problem-solving.
CF-25	Contract completeness	The long project duration had led to the incorporation of detailed contract conditions and contractual procedures to deal with contingencies.
CF-26	Inter-party reciprocity	The previous dealing(s) among project team members enabled them to be more devoted to complete the project.
CF-27		Project team members desired to maintain relationships with the other during the project.

cooperative drivers can be developed by the use of Principal Component Factor Analysis (PCFA) that explores the structure of the inter-relationships among data by defining a set of common underlying constructs, known as factors (Hair *et al.* 1995). Separate dimensions of the structure are then easily identified and interpreted. Against the drivers long-listed in Tables 4.1 and 4.2, the respondents were asked to rate their degree of significance in shaping their contracting behaviours by a Likert scale of 1 (not significant) to 7 (very significant).

The data collected for aggressive and cooperative drivers were subjected to a PCFA to develop their respective taxonomies. As such, two PCFA were performed. The suitability of the data set was first assessed by Kaiser-Meyer-Olkin (KMO) measure of sampling adequacy. The KMO values for the taxonomies of aggressive and cooperative drivers are 0.872 and 0.769 respectively, both are above the threshold requirement of 0.5 (Cheung and Yeung 1998; Cheung *et al.* 2000; Holt 1997). The low significance of the Bartlett test of sphericity also supports the adequacy of the data set to perform PCFA. Furthermore, the eigenvalue-greater-than-1 principle was applied to decide on the number of factors. Factors with the eigenvalue greater than 1 were considered significant, and those with eigenvalue below 1 were discarded. Varimax rotation was applied so as to simplify the factor structures and obtain factor solutions that are easier to interpret. To explain the correlation between aggressive/cooperative drivers and its factor, factor loadings are also shown in Tables 4.3 and 4.4. These loadings give an indication of the extent to which the drivers are influential in forming the factors (Sharma 1996). According to Sharma (1996), the loadings can be obtained by using the following equation:

$$l_{ab} = \frac{W_{ab}}{\hat{s}_b}\sqrt{\lambda_a} \qquad \text{(Eq. 4.1)}$$

where l_{ab} is the loading of the bth driver for the ath factors, W_{ab} is the weight of the bth drivers for the ath factors, λ_a is the eigenvalue of the ath factor, and $\hat{s}_b$ is the standard deviation of the bth driver.

Factor interpretation

As shown in Table 4.3, three factors were extracted for the aggressive drivers. These are: 'Factor 1: unfavourable past experience/ambiguous contract terms'; 'Factor 2: difficulties in performing contract' and 'Factor 3: goal-orientated'. These results are generally supported by previous studies. For example, Luo (2002) suggested that past dealings or experience might influence subsequent dealings between the same contracting parties. Having favourable previous dealings or experience could help contracting parties build up their trusting relationships (Rapoport and Chammah 1965), to appreciate their respective organisational strength and management style (Tallman and Shenkar 1994; Rubin and Brown 1975) and to heighten situational flexibility (Larson 1992; Gulati 1995) Moreover, difficulties in performing contract are one of the drivers to the aggressive moves of contracting parties. Practical difficulties such as the existence of exculpatory and onerous provisions, the change of project content and the low interdependency between contracting parties may bring about ambiguity that creates a breeding ground for shrinking responsibility and shifting blame (Goldberg 1992). As such, contracting parties are more likely to invoke aggressive behaviours. When this happens, aggressive retaliation from the other contracting parties can be expected (Dozzi *et al.* 1996; Hannan and Freeman 1984).

Furthermore, seven factors were extracted for the cooperative drivers. These are: 'Factor 1: openness to contracting parties/contractual settings'; 'Factor 2: good relationships among contracting parties'; 'Factor 3: contract completeness'; 'Factor 4: good teamwork'; 'Factor 5: incentive to risk-sharing'; 'Factor 6: effective communication'; 'Factor 7: desire to maintain relationship'. To motivate cooperative moves, Factor 1 suggests that contracting parties need to be open to each other. Openness is defined as the active involvement within project teams with open/honest communications, exchange of thoughts and feelings and equitable risk sharing. With these, cooperation between contracting parties can be more enduring. Resistance against aggression will also be higher (Crosby and Taylor 1983; Rahtz and Moore 1983).

Table 4.3 Taxonomies of aggressive drivers

Aggressive forces		Factor		
		1	2	3
Factor 1: unfavourable past experience/ambiguous contract terms				
AF-1	The quality of past dealing(s) between project participants was poor (low degree of satisfaction of previous cooperation).	.791	.050	.202
AF-2	The previous dealing(s) were unsuccessful in achieving the goals of the project(s).	.790	.125	.064
AF-15	The unfavourable past cooperation between project participants had caused disputes.	.754	.323	.007
AF-6	The capital necessary for the project operation had been in general insufficient.	.712	.133	.138
AF-14	Not clearly stipulated contract conditions regarding the respective rights, benefits and responsibilities had caused disputes.	.703	.146	.316
AF-4	The actions being taken by competitors/other contracting parties were strongly aggressive.	.660	.263	.263
AF-16	Overly detailed contractual procedures to deal with contingencies had caused disputes.	.659	.493	.060
AF-17	There were many ambiguous terms in the Conditions of Contract used.	.620	.521	−.050
Factor 2: difficulties in performing contract				
AF-11	Project complexity made your project team to gain an advantage over the other parties.	−.107	.767	.222
AF-13	The existence of exculpatory provisions had caused disputes in this project.	.591	.619	.121
AF-3	There had been great changes in the project content, due to environmental issues arising during the course of construction.	.417	.613	.098
AF-10	The contract conditions were onerous and the performance specifications were harsh.	.491	.537	.091
AF-12	Low interdependency between the project participants had led to your party more likely taking advantage over the others.	.390	.517	.334
Factor 3: goal-orientated				
AF-8	Having fair expectations of future profits and rewards would make your project team more likely to gain an advantage over the other parties.	.233	.220	.832
AF-9	Perception to aggressive actions of competitor/other contracting parties would more likely make your project team adopt an aggressive stance.	.227	.165	.792
AF-5	Dealing with the issues that can increase profitability would increase the competitive pressure of your project team.	.450	−.081	.702
AF-7	Your project team would become more active to deal with the issues that can benefit to achieve your goal.	−.169	.132	.694

Factor 2 highlights the importance of having good relationships among contracting parties. Good relationship between project participants is the foundation of cooperation (Baker *et al.* 1983; Cheung *et al.* 2004; Hartman 1993). Contracting parties, in particular, with previous cooperative experience, would reciprocate in subsequent dealings (Katz and Kahn 1978). Blau (1964) and Luo (2002) recognise that current behaviour is a response to previous behaviour of the others, thus past dealings are having influence on the cooperation or otherwise between the same contracting parties.

As suggested by Turner (2004), contracts should aim at aligning a contractor's objectives with those of the clients. An appropriate contractual arrangement could attenuate the leeway for opportunism and prohibit moral hazards in a cooperative relationship (Hackett 1993). Factor 3, is a multidimensional concept and concerns not only term specificity but also contingency adaptability (Luo 2002). It also serves as a framework in governing ongoing cooperation and facilitating the contribution of cooperation to performance (Luo 2002; Williamson 1996).

Successful project delivery requires the concerted effort of all contracting parties in executing and supervising the construction processes (Cheung *et al.* 2004). The stronger the teamwork, the greater the cooperative drivers the contracting parties possess. Factor 4 reveals that the capability of maintaining good teamwork spirits is one of the drivers to motivate contracting parties to adopt cooperative moves. Team effectiveness has been identified as a reliable indicator of the degree of cooperation of the contracting parties (Crane *et al.* 1999; Luo 2002).

Factor 5 consists of those cooperative drivers that can be collectively described as the incentive to risk-sharing/problem-solving. As suggested by McKim (1992) and Jannadia *et al.* (2000), contracting parties would be less likely to adopt cooperative moves, or may even adopt aggressive moves, if the contract conditions are strict, harsh and onerous. Risk aversion attitude is the greatest obstacle against the use of cooperative contracting (Dozzi *et al.* 1996; Hartman 1993). Similarly, the degree of cooperation between contracting parties could be indicated by whether there was mutual consultation concerning issues under uncertain conditions. That means trust among project participants enable problems to be solved more effectively (Luo 2002).

Factor 6 is essential for the initiation of cooperative moves (Lewis and Weigert 1985). It promotes open and efficient information exchange/ interpretation between contracting parties (Rubin and Brown 1975). Communication channels to identify project progress and problems enables the prompt resolution of these issues in a cooperative fashion and hence maintains a climate of mutual cooperation among contracting parties (Turner 2004). In sum, the availability of information could affect the attitudes of contracting parties (Bibb and Nowak 1994) Contracting parties would more likely be cooperative if: (a) there is satisfying past cooperation among them; (b) there is no constraint in getting information from each other; and (c) they have considerable experience in handling projects of similar nature.

Moving on to Factor 7, with the desire to maintain future relationships, contracting parties would be more likely to reciprocate their cooperative endeavour with each other. This would imply that past, present and future relationships are linked to the cooperative moves of contracting parties.

Table 4.4 Taxonomies of cooperative drivers

Cooperative drivers		*Factors*						
		1	2	3	4	5	6	7
Factor 1: openness of contracting parties/contractual settings								
CF-10	A high degree of involvement within project by team members.	.842	.180	−.029	.122	−.085	.029	−.035
CF-9	Project team member had open and honest communications.	.814	.149	.004	.193	−.111	.169	−.016
CF-12	The contract design encouraged project team member to achieve common objectives in a rational manner.	.805	.147	−.007	.037	.025	−.311	.117
CF-15	Project team members had plenty of experience in handling project(s) of similar nature.	.767	.196	−.262	−.028	.035	.133	.153
CF-18	The contract provisions apportioned risks equitably between contracting parties.	.760	.113	.115	−.094	−.073	−.035	.066
CF-19	The contract provisions and specifications provided incentive for risk sharing between the contracting parties.	.736	−.034	.143	.043	.325	.073	−.121
CF-13	There were no constraints in getting information from other project team members.	.715	.159	.110	.066	−.137	−.110	−.053
CF-8	Project team members were willing to share thoughts and feelings with each other.	.683	.239	.199	.052	.071	.005	−.023
CF-14	The project team members had mutual goals.	.682	−.090	−.129	.317	−.012	−.078	.219
CF-5	A good working relationship from previous dealing(s) among project team members.	.553	.416	.055	.504	.046	−.057	−.121

Table 4.4 continued

Cooperative drivers		Factors						
		1	2	3	4	5	6	7

Factor 2: good relationships among contracting parties

CF-17	The involvement of your project team members had been highly voluntary.	.315	.705	.203	−.126	−.024	−.040	.294
CF-6	A good personal relationship among project team members.	.126	.633	.122	.371	−.048	.012	.325
CF-16	Good information exchange from previous dealing(s) among project team members.	.288	.618	−.200	−.031	.163	−.310	−.017
CF-2	Previous dealing(s) among project team members reinforced confidence in working with each other.	.271	.591	.030	.355	.033	.202	−.004
CF-11	Your project team demonstrated open commitment within project team members.	.149	.490	.363	−.086	−.315	.227	−.380

Factor 3: contract completeness

CF-25	The long project duration had led to the incorporation of detailed contract conditions and contractual procedures to deal with contingencies.	.035	.125	.867	−.038	−.040	.181	.009
CF-26	The previous dealing(s) among project team members enabled them to be more devoted to complete the project.	.084	−.002	.819	.083	.021	.170	.220
CF-23	Provision of a third party who was paid jointly by the contracting parties (e.g. an experienced mediator) enabled effective dispute resolution.	.059	.024	.569	.306	.164	−.121	−.451

Factor 4: good teamwork

CF-4	Teamwork spirit among project team members facilitated project progress effectively.	.072	.050	.052	.795	.098	.147	.155
CF-1	Teamwork spirit facilitated dispute resolution effectively.	.516	.330	.043	.541	−.219	−.034	−.154

Table 4.4 continued

Cooperative drivers				Factors				
		1	2	3	4	5	6	7
Factor 5: incentive to risk sharing/problem solving								
CF-20	Provision of tangible reward (e.g. bonus, a gain share fund, etc.) enabled the project team members to generate more incentive to save cost.	.115	−.064	.092	.387	−.723	.128	−.044
CF-22	Guidelines of handling various unanticipated contingencies had been incorporated in the contract in most cases.	−.013	−.053	.002	.133	.690	.249	−.012
CF-24	Mutual consultation between contracting parties enabled effective problem solving.	.111	.030	.117	.302	.561	.219	−.213
CF-21	A modest contract sum would make project team members less risk averse.	−.106	.059	.415	−.046	.445	.232	.353
Factor 6: effective communication								
CF-3	Effective communication from previous dealing(s) among project team members.	−.036	.015	.151	−.024	.110	.816	−.014
CF-7	Trust developed among project team members facilitated effective project progress.	−.017	−.032	.146	.202	.208	.753	.066
Factor 7: desire to maintain relationship								
CF-27	Project team members desired to maintain relationships with the others during the project.	.149	.258	.182	.154	−.056	.017	.754

Stage 3: relative importance rankings of the aggressive and cooperative drivers

Examining the factors for aggressive and cooperative drivers as identified in the two PCFAs enables the understanding of aggressive-cooperative nature of CCB in a more amenable and logical manner. The significance of these drivers may vary between construction practitioners with different contracting behaviours inclinations. In this connection, ranking the relative importance of these factors for a particular contracting behaviour will help the contracting parties to adopt appropriate moves. Therefore, this stage aims at ranking the identified factors by their relative importance in the light of the contracting behaviours self-reported by the respondents. As described previously, the second section of the questionnaire includes an assessment by the respondent of the degree of aggressiveness and cooperativeness of the CCB. A majority of respondents described their contracting behaviours as neutral, accommodative and cooperative for their recent completed project. Only four of the 100 respondents reported confrontational contracting behaviours being adopted (see Figure 4.2).

Based on the factors previously extracted by the PCFAs, factor scores can be computed for each of these self-reported contracting behaviours. With these factor scores, the degree of significance of each extracted factor can be revealed in the views of the respondents with confrontational, neutral, accommodative and cooperative contracting behaviours. In this connection, the entire data set was divided into four sub-groups. Each of these subgroups represents the data collected from the respondents with confrontational, neutral, accommodative and cooperative contracting behaviours. For each sub-group, 10 factor scores (which are developed from the 3 and 7 factors of aggressive and cooperative forces respectively) can be calculated by averaging the mean score of the attributes of each factor. For example, Factor 1 consists of eight attributes (AF-1, AF-2, AF-15, AF-6, AF-14, AF-4, AF16 and AF-17) (see Table 4.3). Its factor scale can be computed by averaging the mean scores of these attributes with the following formula:

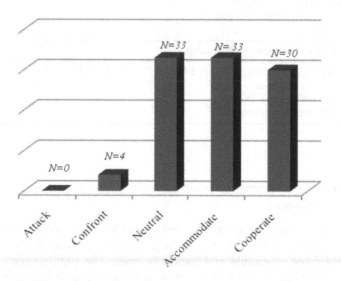

Figure 4.2 Self-reported contracting behaviours of the respondent (Total N =100)

$$FS_{ij} = \frac{\sum_{k=1}^{n} M_{ijk}}{n} \qquad\qquad \text{(Eq. 4.2)}$$

where FS_{ij} is the factor scale of factor i in the jth sub-group, M_{ijk} is the mean score of the kth attribute of factor i in the jth sub-group, and n is the number of attributes for each factor.

With Equation 4.2, the 10 factor scales were calculated and ranked for each of the five sub-groups of reported contracting behaviours (see Table 4.5).

Table 4.5 Rankings of factor scale for different self-reported contracting behaviours

Taxonomies	Reported contracting behaviours (Total N = 100): factor scores (Ranking)				
	Attack (N=0)	Confront (N=4)	Neutral (N=33)	Accommodate (N=33)	Cooperate (N=30)
Aggressive drivers					
Factor 1: Unfavourable past experience/ambiguous contract terms	—	4.438 (2)	3.947 (3)	3.644 (3)	2.800 (3)
Factor 2: Difficulties in performing contract	—	4.550 (1)	4.000 (2)	3.661 (2)	3.260 (2)
Factor 3: Goal-oriented	—	4.188 (3)	4.341 (1)	4.530 (1)	4.333 (1)
Cooperative drivers					
Factor 1: Openness of contracting parties/contractual settings	—	4.000 (5)	4.185 (3)	4.803 (1)	5.407 (1)
Factor 2: Good relationships among contracting parties	—	4.300 (2)	4.442 (1)	4.661 (3)	5.280 (3)
Factor 3: Contract completeness	—	4.083 (4)	3.859 (6)	4.394 (4)	3.811 (6)
Factor 4: Good teamwork	—	4.125 (3)	4.258 (2)	4.773 (2)	5.300 (2)
Factor 5: Incentive to risk-sharing/problem-solving	—	3.938 (6)	4.136 (4)	3.955 (6)	4.283 (5)
Factor 6: Effective communication	—	3.125 (7)	3.439 (7)	3.136 (7)	2.950 (7)
Factor 7: Desire to maintain relationship	—	5.000 (1)	4.121 (5)	4.333 (5)	4.667 (4)

Summary

This study identified the aggressive and cooperative drivers in construction contracting. With the use of the questionnaire survey, the respondents were asked to rate their own aggressive and cooperative drivers and to address the degrees of aggressiveness and cooperativeness of their contracting behaviours based on their recent completed construction project. As such, the perceived behaviour of the respondents against actual contracting behaviour of the reported projects can be assessed. As shown in Table 4.5, it can be noted that the majority of the respondents did not consider their contracting behaviour as aggressive despite the general view that construction contracting is confrontational. Thirty per cent of the respondents in fact rated themselves as cooperative and another 33 per cent rated as accommodating. For these two groups of respondents, their cooperative taxonomies scores are higher than their aggressive taxonomies scores. These support their self-evaluations. The most important cooperative taxonomy is openness of contracting parties that covers ten drivers (Table 4.4 refers). In essence, this is a pragmatic and effective approach to tackle construction problems that are characterised by multi-party involvement, interdependency and are conflict laden. It also makes good sense that cooperation is supported by good teamwork (second most important cooperative taxonomy) and good relationships among contracting parties (third most important cooperative taxonomy). With regard to aggressive drivers, 'goal oriented' was ranked as the most important group of drivers. This reflects the downside of over-emphasising self-interest in construction contracting. Whilst this may be attitudinal, the second most important aggressive driver group is related to what happens during construction. Perhaps this group of drivers can have better chances to be reduced by cooperative drivers. Unfavourable past experience was ranked in the third place. This cognitive factor can also be corrected by cooperative moves, although the initial barrier can be substantial. As for the 'neutral' group of respondents, the rankings of the aggressive taxonomies are similar to the 'cooperative' group. In the case of cooperative drivers, the first top three ranked are the same except the top most and the third most are in reverse order. This suggests that the neutral group is more likely to behave in ways that reflect the strength of their relationships with their contracting parties. Only four respondents considered their contracting behaviour confrontational. The number of responses is too small for any form of generalisation; notwithstanding that construction is generally identified as a confrontational industry with many projects ending up with significant disputes. It is of equal truth that cooperative contracting can bring substantial benefits. Thus it is reasonable to assume that in every construction project, both cooperative and aggressive drivers that affect the behaviour of the contracting parties co-exist. The findings of this study suggest that construction projects may not be inevitably adversarial. Most respondents of the study identified themselves as non-confrontational. Nonetheless, this does not help in explaining why projects remain quite likely to end in dispute. This invites further study in this respect. Contracts with equitable risk allocation supported by open discussion of problems provide the platform for team building whereby relationships among the contracting parties can be maintained. The ability to build effective teams is an important soft power of the construction contracting organisation.

Acknowledgements

Special thanks to Miss On Ki Chiu for collecting data for the study. The content of this chapter has been published in Volume 27(4) of the *International Journal of Project Management* and is used with the permission from Elsevier.

References

Amit, R., Domowitz, I. and Fershtman, C. (1988) Thinking one step ahead: the use of conjectures in competitor analysis. *Strategic Management Journal*, 9, 431–42.

Associated General Contractors of America (AGCA) (1991) *Partnering: A Concept for Success*. Washington, DC: AGCA.

Baker, B. N., Murphy, D. C. and Fisher, D. (1983) *Factors Affecting Project Success: Project Management Handbook*. New York: Van Nostrand Reinhold,.

Bayliss, R., Cheung, S. O., Suen, C. H., Wong, S. P. (2004) Effective partnering tools in construction: A case study on MTRC TKE Contract 604 in Hong Kong. *International Journal of Project Management*, 22, 253–63.

Bennett, J. and Jayes, S. (1995) *Trusting the Team: The Best Practice Guide to Partnering in Construction*. Centre for Strategic Studies in Construction/Reading Construction Forum.

Bibb, L., Nowak, A. (1994) Attitudes as catastrophes: from dimensions to categories with increasing involvement, in R. R. Vallacher and A. Nowak (eds), *Dynamical Systems in Social Psychology*. San Diego, CA: Academic Press, 219–49.

Black, C., Akintoye, A., and Fitegerald, E. (1999) An analysis of success factors and benefits of partnering in construction. *International Journal of Project Management*, 18, 423–34.

Blau, P. M. (1964) *Exchange and Power in Social Life*. New York: J. Wiley.

Bramble, B. B. and Cipollini, M. D. (1995) *Resolution of Disputes to Avoid Construction Claims*. Transportation Research Board, Synthesis of Highway Practice 24. Washington, DC: National Academy Press.

Byrnes, J. D. (2002) *Before Conflict: Preventing Aggressive Behavior*. Lanham, MD: Scarecrow Press.

Cheung, S. O. and Suen, C. H. (2002) A multi-attribute utility model for dispute resolution strategy selection. *Construction Management and Economics*, 20, 557–68.

Cheung, S. O., Suen, C. H. and Cheung, K. W. (2003) An automated partnering monitoring system – partnering temperature index. *Automation in Construction*, 12, 331–45.

Cheung, S. O., Suen, C. H., Cheung, K.W. (2004) PPMS: a web-based construction project performance monitoring system. *Automation in Construction*, 13(3), 361–76.

Cheung, S. O., Tam, C. M., Ndekugri, I. and Harris, F. C. (2000) Factor affecting clients' project dispute resolution satisfaction in Hong Kong. *Construction Management and Economics*, 18, 281–94.

Cheung, S.O. and Yeung, Y.W. (1998) The effectiveness of the dispute resolution advisor system: a critical appraisal. *The International Journal of Project Management*, 16(6), 367–74.

Construction Industry Institute, Australia (CIIA) (1996) *Partnering: Models for Success*. Brisbane: CIIA.

Construction Industry Review Committee (CIRC) (2001) *Construct for Excellence*. Hong Kong: Construction Industry Review Committee.

Crane, T. G., Felder, J. P., Thompson, P. J., Thompson, M. G. and Sanders, S. R. (1999) Partnering parameters. *Journal of Management in Engineering*, 15(2), 37–42.

Crosby, A., Taylor, R. (1983) Psychological commitment and its effects on post-decision evaluation and preference stability among voters. *Journal of Consumer Research*, 9, 413–31.

DeVilbiss, C. E. and Leonard, P. (2000) Partnering is the foundation of a learning organization. *Journal of Management in Engineering*, 16(4), 47–57.

Dozzi, P., Hartman, F., Tidsbury, N. and Ashrafi, R. (1996) More-stable owner-contractor relationship. *Journal of Construction Engineering and Management*, 155(1): 31–35.

Dutton, J. and Duncan, R. (1987) The creation of momentum for change through the process of strategic issue diagnosis. *Strategic Management Journal*, 8, 279–95.

Egan, J. (1998) *Rethinking Construction*. London: Department of the Environment, Transport and the Regions.

Fenn, P., Lowe, D. and Speck, C. (1997) Conflict and dispute in construction. *Construction Management and Economics*, 15(6), 513–18.

Goldberg, V. P. (1992) The past is the past – Or is it? The use of retrospective accounts as indicators of past strategy. *Academy of Management Journal*, 35, 848–60.

Gulati, R. (1995) Does familiarity breed trust? The implications of repeated ties for contractual choice in alliances. *Academy of Management Journal*, 38(1), 85–112.

Hackett, S.C. (1993) Incomplete contracting: a laboratory experimental analysis. *Economic Inquiry*, 31, 274–97.

Hair, A., Tatham, R. L., Black, W. C. (1995) *Multivariate Data Analysis* (5th edn). Englewood Cliffs, NJ: Prentice-Hall.

Hannan, M. and Freeman, J. (1984) Structural inertia and organizational change. *American Journal of Sociology*, 49, 149–64.

Harmon, K. M. (2001) Pseudo arbitration clauses in New York City construction contracts. *Construction Briefings*, 7.

Harmon, K. M. (2003) Conflicts between owner and contractors: proposed intervention process. *Journal of Management in Engineering*, 19(3), 121–25.

Hartman, F. T. (1993) Construction Dispute Resolution through an Improved Contracting Process in the Canadian Context, PhD Thesis, Loughborough University of Technology.

Holt, G. (1997) Construction research questionnaire and attitude measurement: relative index or mean. *Journal of Construction Procurement*, 3(2), 88–94.

Jannadia, M. O., Assaf, S., Bubshait, A. A. and Naji, A. (2000) Contractual methods for dispute avoidance and resolution (DAR). *International Journal of Project Management*, 18(6), 41–49.

Katz, D. and Kahn, R. L. (1978) *The Social Psychology of Organization* (2nd edn). New York: Wiley.

Kumaraswamy, M. M. and Matthews, J. D. (2000) Improved subcontractor selection employing partnering principles. *Journal of Management in Engineering*, 16(3), 47.

Larson, A. (1992) Network dyads in entrepreneurial settings: a study of the governance of exchange relationships. *Administrative Science Quarterly*, 37, 76–104.

Latham, M. (1994) Constructing the team: final report by Sir Michael Latham, Joint Review of Procurement and Contractual Arrangements in the United Kingdom Construction Industry. London: HMSO.

Lewis, J. D. and Weigert, A. (1985) Trust as a social reality. *Social Forces*, 63, 967–85.

Li, H., Cheng, H. and Love, P. (2000) Partnering research in construction. *Engineering, Construction and Architectural Management*, 7, 76–92.

Luo, Y. (2002) Contract, cooperation and performance in international joint ventures. *Strategic Management Journal*, 23, 903–19.

Macmillan, C. and Jennifer, A. (1990) Resource cooptation via social contracting: resource acquisition strategies for new ventures. *Strategic Management Journal*, 11, 79–92.

Mamer, J. and McCardle, K. (1987) Uncertainty, competition and adoption of new technology. *Management Science*, 33, 161–77.

McKim, R. A. (1992) Risk behavior of contractors: a Canadian study. *Project Management Journal*, 23(3), 51–55.

Monahan, G. (1987) The structuring of equilibria in market share attraction models. *Management Science*, 33, 228–43.

Parkhe, A. (1993) Strategic alliance structuring: a game theoretic and transaction cost examination of interfirm cooperation. *Academy of Management Journal*, 36, 794–829.

Pinnell, S. (1999) Partnering and the management of construction disputes. *Dispute Resolution Journal*, 54(1), 16–22.

Rahtz, R. and Moore, L. (1983) Product class involvement and purchase intent. *Psychology and Marketing*, 47, 68–78.

Rapoport, A. and Chammah, A. M. (1965) *Prisoner's Dilemma*. Ann Arbor: University of Michigan Press.

Rosch, E. (1988) Principles of categorisation, in A. Collins and E. Smith (eds), *Readings in Cognitive Science*. Los Altos, CA: Morgan Kaufmann, 312–322.

Rubin, J. Z., Brown, B. R. (1975) *The Social Psychology of Bargaining and Negotiation*. New York: Academic Press.

Sanvido, V., Grobler, F., Parfitt, K., Guvenis, M. and Coyle M. (1992) Critical success factors for construction projects. *Journal of Construction Engineering and Management*, 118(1), 94–111.

Sharma, S. C. (1996) *Applied Multivariate Techniques*. New York: Wiley.

Swedberg, R. (1987) Economic sociology: past and present. *Current Strategic*, 5(1), 1–21.

Tallman, S. and Shenkar, O. (1994) A managerial decision model of international cooperative venture formation. *Journal of International Business Studies*, 25, 91–114.

Turner, J. R. (2004) Farsighted project contract management: Incomplete in its entirety. *Construction Management and Economics*, 22, 75–83.

Tushman, M. L., Romanelli, E. (1985) Organizational evolution: a metamorphosis model of convergence and reorientation. *Research in Organizational Behavior*, 7(1), 171–222.

Williamson, O. E. (1996) *The Mechanisms of Governance*. New York: Oxford University Press.

Yiu, T. W. (2007) Forces to foster co-operative contracting in construction projects. *Australasian Dispute Resolution Journal*, 18(2), 113–18.

Yiu, T. W. and Cheung, S. O. (2006) A catastrophe model of construction conflict behaviour. *Building and Environment*, 40(1), 438–47.

Zack, J. G. (1995) Practical dispute management. *Cost Engineering*, 37(12), 55.

Section C

Becoming a learning organisation

5 The concept of organisational learning in construction and its effect on project performance

Peter Shek Pui Wong and Sai On Cheung

Introduction

Construction contracting organisations in this book refer to the organisations collaborating along the construction supply chain. This includes the developers, architects, engineering and surveying consultants, main contractors, sub-contractors and the suppliers. Construction contracting organisations have been criticised as incapable of solving unprecedented problems, grasping unanticipated opportunities and adapting to the dynamic business environment (Love *et al.* 2000). Furthermore, they have been described as inflexible and slow to respond to the escalating and changing demands of customers (Holt *et al.* 2000; Love *et al.* 2004). To these ends, researchers emphasised that construction contracting organisations should embrace performance improvement (PER) as one of the project goals (Holt *et al.* 2000; Jashapara 2003; Kululanga *et al.* 1999; Kumaraswamy 1998; Love *et al.* 2000). Attaining performance improvement has become one of the major strategic foci of construction contracting organisations in the past decade (Holt *et al.* 2000). However, industry reviews conducted in different countries report that performance of the construction industry is generally unsatisfactory (Egan 1998; CIRC 2001). Under-performance is manifested by erosion of productivity, reduction in profitability and mounting inter-firm adversarial relationships within the supply chain (Love *et al.* 2004). Moreover, this problem becomes more acute as if one of the organisations in the supply chain fails to perform the whole team may lose out in this ever-intensifying competitive construction market (Wong and Cheung 2005a). Hence, there are good reasons to believe that performance improvement cannot be easily achieved.

In a typical construction project, a performance improvement (PER) process can be presented by a Plan-Do-Study-Respond cycle as shown in Figure 5.1 (Law and Chuah 2004). The cycle starts with 'Plan': a construction contracting organisation should first formulate a detailed plan that includes steps to be taken to achieve the anticipated performance goals. This plan is then implemented in the 'do' stage. In this connection, various sets of performance metrics have been suggested (Zipf 1998; Chaaya and

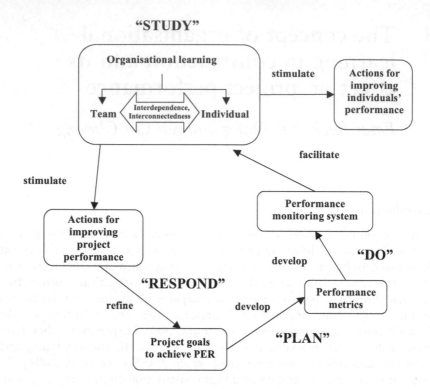

Figure 5.1 A CI framework for construction projects

Jaafari 2001). For example, Zipf measured construction contracting organisations' performance in terms of time and cost required for completing the assigned construction work. Likewise, Chaaya and Jaafari (2001) developed a set of metrics that evaluate construction contracting organisations' performance in terms of their achievement of the pre-agreed time, cost and quality targets. The achievement or otherwise of these metrics are used as measurements of project performance. Furthermore, feedback loop is always critical for a management cycle. In this regard, some researchers advocated performance records as a source of feedbacks from which lessons can be learnt (Crawford and Bryne 2003; Ozorhon *et al.* 2005). A chain of studies on Performance Measurement System (described as the PMS thereafter) were noted (Al-JiBouri 2003; Crawford and Bryne 2003). For most of these systems, the primary aim is to ensure proper periodic recording of information reflecting project performance. Moreover, some researchers also described these systems as an invaluable 'source of knowledge' for the construction contracting organisations (Ozorhon *et al.* 2005). In this respect, the implementation of PMS can help in formalising the knowledge acquisition process (Crawford and Bryne 2003).

Nonetheless, performance improvement could only be achieved if construction contracting organisations can respond appropriately after acquiring knowledge derived from the PMS (Al-JiBouri 2003). In this regard, the process of applying the imbibed knowledge for performance improvement has been coined as organisational learning (OL) (Kululanga *et al.* 1999, 2002; Love *et al.* 2004; Ozorhon *et al.* 2005). In other words, the ability of construction contracting organisations to practice OL is critical in completing the loop for performance improvement (Ozorhon *et al.* 2005).

Meanings of organisational learning

Organisational learning (OL) is a developing and emerging research topic in construction. Previous research mainly aimed to import OL to construction based on successful experience obtained from other fields (Wong and Cheung 2008). These studies largely adopted Argyris and Schön's (1978) action-based theory to explain learning (Kululanga *et al.* 1999; Love *et al.* 2000; Jashapara 2003; Murray and Chapman 2003; Wong and Cheung 2008). Indeed, Argyris and Schön's OL definition is recognised as the first deliberation of OL theory (Gherardi 2002). They conceptualised OL as a process of detection and correction of errors found from both internal and external environments (Argyris and Schön 1978). Since then, their efforts have been advanced further by a number of researchers (Bennis and Nanus 1985; Duncan and Weiss 1979; Fiol and Lyles 1985). Duncan and Weiss reviewed the definitions of learning for organisations and argued that OL is not merely a process of 'detection and correction of errors' but a behavioural change that shortens the gap between actual and expected performance outcomes. Bennis and Nanus (1985) extended Argyris and Schön's concept by defining OL as a series of processes of acquiring knowledge from past actions, followed by the transformation of knowledge to behaviours, tools and strategies that would facilitate future improvement actions. A similar definition was used by Fiol and Lyles (1985) who defined OL as a 'process of knowledge and understanding for past actions and future improvement actions'. The notions of OL in construction were also expounded from these classical definitions (Franco *et al.* 2004; Huemer and Östergren 2000; Jashapara 2003; Love *et al.* 2000, 2004). This study benefited from this wealth of research and defines OL of the construction contracting organisations as a process of applying the imbibed knowledge for performance improvement. As such, the knowledge imbibed is captured by construction contracting organisations for improvement actions as and when they become necessary (Kululanga *et al.* 1999). Figure 5.2 depicts how researchers conceptualise OL in construction.

Generally, OL in construction involves a four-stage process encompassing knowledge acquisition, knowledge integration, organisational memory

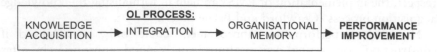

Figure 5.2 Organisational learning in construction (modified from Ozorhon *et al.* 2005)

and improvement actions (Love and Josephson 2004, Ozorhon *et al.* 2005).

The process starts from knowledge acquisition, which is triggered both intrinsically and extrinsically (Kululanga and MaCaffer 2001). Intrinsically, construction contracting organisations acquire knowledge from their experience gained by performing repetitive tasks (Kululanga *et al.* 2001; Pawlowsky 2001) and internal benchmarking (Kululanga and MaCaffer 2001). Extrinsically, construction contracting organisations acquire knowledge from other firms (Jashapara 2003, Murray and Chapman 2003) and their adaptation to changes (Kululanga and MaCaffer 2001). In this study, knowledge that can be acquired intrinsically is described as intrinsic knowledge and knowledge that can be acquired extrinsically is described as extrinsic knowledge. At the stage of knowledge integration, the organisation integrates both intrinsic and extrinsic knowledge into the decision(s) of performance rectification. The respective decisions of performance rectification then become part of the organisational memory. Organisational memory can be defined as the knowledge acquired from the past that can be brought to incorporate present activities (Ozorhon *et al.* 2005). To accomplish performance improvement, the organisational memory has to be transformed to appropriate improvement actions (Baltrusch 2001; Huber 1991; Ozorhon *et al.* 2005).

However, previous studies identified ineffective knowledge acquisition and integration as a major obstacle to the successful practice of OL in construction (Cayes 1998; Love *et al.* 2004; Oyegoke 2006). Furthermore, there has been a lack of infrastructures developed to facilitate knowledge acquisition and integration (Ford *et al.* 2000). Love *et al.* (2000) quoted the study of Garvin (1993) and identified that a learning organisation in the construction industry should be skilled at acquiring knowledge from its own experiences and those of other collaborating firms, from past records and through transferring knowledge quickly and efficiently within. Nonetheless, they discovered that there exists a lack of guidelines for knowledge acquisition and sharing.

Although organisations can maintain or improve their performance based on their experience, some case studies have also indicated that not all organisations can capture extrinsic knowledge effectively from other construction contracting organisations (Bröchner *et al.* 2002; Goulding and Alshawi 2004; Kale and Arditi 1999).

Previous research findings about organisational learning in construction

Based on the results collected from interviewing project managers in the UK, Bresnen *et al.* identified the mobility of employees as a major barrier to effective knowledge capture and transfer within construction contracting organisations. To tackle this problem, they suggested developing a knowledge management approach to facilitate capturing, recording and sharing of experiences among staffs (Bresnen *et al.* 2003).

Nevertheless, in a case study that evaluated the contractors' learning capability in the UK, Davey *et al.* (2004) found that construction contracting organisations are incapable of integrating the knowledge acquired from their routines. Furthermore, they identified the lack of a system that facilitates knowledge acquisition.

Contrary to the above findings, Ozorhon *et al.* (2005) also based on case studies and reported that Turkish contractors have made use of several knowledge sources and mechanisms to acquire knowledge. Nonetheless, they were 'weak in exploitation of organisational memory, especially at the strategic decision-making stage'.

Given that the construction industry is project-based, discontinuous and fragmented, Ozorhon *et al.* (2005) pinpointed the importance in having an organisational memory. In particular, some studies examined the capability of organisations in transferring knowledge gained from one project to another (Bresnen *et al.* 2003; Dorée and Holmen 2004; Yashiro 2001). Bresnen *et al.* (2003) argued that construction contracting organisations generally lack a systematic approach to store and administer the knowledge gained from multiple projects. Along a similar line of research focus, Dorée and Holmen (2004) identified the inability of organisations to transfer their knowledge gained from one project to another. Yashiro (2001) identified that an organisation's project-to-project knowledge transfer depends on the cooperativeness among the collaborating firms. Nonetheless, his case studies reported that construction contracting organisations often fail to transfer knowledge from one project to another project due to the unsupportive attitudes of their working partners. In sum, studies about the practice of OL in construction were mainly based on literature reviews and case studies.

The research problem

The Plan-Do-Study-Respond cycle as shown in Figure 5.1 shows how PER can be achieved by construction contracting organisations under an OL framework. The significance of developing OL as part of the construction contracting organisations' routines has attracted the attention of a number of researchers (Jashapara 2003; Kululanga *et al.* 1999; Murray and Chapman 2003). Kululanga *et al.* (1999) pinpointed that OL 'offers avenues to bring about a continuous improvement agenda' in the construction operational process. Murray and Chapman (2003) stressed that facilitating OL process is a fruitful mission that construction practitioners should aim to achieve. Generally, the implementation of OL has been advocated as one of the key constructs for PER (Jashapara 2003).

An assortment of studies about OL was reported in this connection (Jashapara 2003; Kululanga *et al.* 1999, 2002; Love *et al.* 2000; Love and Josephson 2004; Murray and Chapman 2003; Wong and Cheung 2008). One of the research foci of these studies was to explore the learning types depicted by the organisations (Kululanga *et al.* 1999, 2002; Wong and Cheung 2008). It has been suggested that an organisation's learning type determines their disposition of knowledge acquisition and transformation for improvement actions (Kululanga *et al.* 1999; Wong and Cheung 2008). Jashapara (2003) applied structural equation modelling (SEM) to confirm empirically the relationships between OL types and contractors' performance. It is noted that performance as mentioned in his study was focused at individuals' level (i.e. was gauged by the performance of an individual working in construction contracting organisation). Despite individual and project performance being interdependent and interconnected, they are gauged in different perspectives (Law and Chuah 2004). In order to bring about PER, construction contracting organisations should learn and perform in order to achieve the common project goals (see Figure 5.1) (Crawford and Bryne 2003). As such, project performance concerns the entire project team's achievements on project efficiency and effectiveness (Crawford and Bryne 2003; Drucker 1974; Mintzberg 1989). Moreover, studies in organisational learning identified the contingent effect of different learning types on outcomes (Murray and Chapman 2003; Wong and Cheung 2008). These studies affirm the proposition that the practice of different OL types affects the attainment of performance improvement (PER) (Law and Chuah 2003, 2004, Ozorhon *et al.* 2005).

A better understanding of the interrelationship among OL types and PER shall provide valuable insights for management to devise ways and means to enhance project success. This study seeks to verify and examine the relationships between practising different OL types and PER in construction projects. To achieve the research objective, the following methodologies were developed:

- First, attributes for identifying the practice of OL types by the construction contracting organisations were reviewed.
- Second, metrics for measuring PER were also developed.
- The above reviews underpin the conceptual model of this study.
- Based on the conceptual model, a questionnaire survey was developed and administered to measure the extent of practicing various OL types by the construction contracting organisations and PER in their respective construction projects.

Types of OL

OL type can be defined as the manner in which an organisation applies the imbibed knowledge for performance improvement (Kululanga *et al.* 1999). Various OL types were identified in previous studies (Argyris and Schön 1978; Easterby-Smith *et al.* 1991; Francis 1997; Hayes and Allinson 1998; Kurtyka 2003). Argyris and Schön (1978) identified that organisations mainly exhibit three major types of learning: single-loop learning, double-loop learning and deutero-learning. This taxonomy is subsequently used by Kululanga *et al.* (1999) to investigate the practice of contractors' learning in the United Kingdom. Murray and Chapman (2003) suggested two principal OL types: adaptive learning and generative learning that match well with the definitions of single-loop and double-loop learning respectively. Jashapara (2003) identified behavioural learning and cognitive learning as the major project learning types of construction contracting organisations. Behavioural learning can be viewed as 'new responses or actions based on existing interpretations'. Cognitive learning refers to the continuous review and modification of ways of working for performance improvements. He further described cognitive learning and behavioural learning as single-loop and double-loop learning respectively. Notwithstanding the difference in terminology, many OL taxonomies in construction stem from the work of Argyris and Schön (1978) (Holt *et al.* 2000; Kululanga *et al.* 1999, 2002; Love *et al.* 2000; Jashapara 2003; Wong and Cheung 2008). This book also employs this taxonomy.

Single-loop learning (SL)

Single-loop learning (SL) refers to the alteration of actions taken based on the divergence or mismatch between the predetermined objectives and actual happenings (Argyris and Schön 1978). This involves recognition and correction of errors in actions to ensure the achievement of the anticipated outcomes (Jashapara 2003). SL can be described as a learning process without scrutinising the underlying reasons that had led to the divergence between the predetermined objectives and the obtained outcomes (Wong *et al.* 2009). An example that explains the concept of SL is of a water heater that starts reheating when a lower than pre-set water temperature is detected. As such, SL is a simple fault detecting and correcting reaction.

Previous studies suggested that the practice of SL can be identified by the following responses: (i) Working (and considering corrective actions if required) under a clearly identified project goal (SL1); and (ii) referring to the firm's past experience to interpret the performance feedback (SL2) (see Table 5.1).

Table 5.1 Attributes for identifying OL types

Practice of OL in terms of:	Kululanga et al. 1999	Pavlowsky 2001	Jashapara 2003	Kurtyka 2003	Murray and Chapman 2003	Love and Josephson 2004	Visser 2004
Single-loop learning (SL)							
1. Working (and considering corrective actions if required) under a clearly identified project goal (SL1)	*	*	*	*	*		
2. Referring to the firm's past experience to interpret the performance feedback (SL2)	*	*	*		*		
Double-loop learning (DL)							
1. Identifying the root of the problem before taking improvement action (DL1)	*	*	*	*	*		
2. Seeking and adopting new management and working approaches through evaluation of current practice (DL2)	*	*	*		*	*	
Deutero-learning (DeuL)							
1. Collecting and focusing on information which reflects the need to improve (DeuL1)	*	*			*	*	*
2. Systematising the communication channels among staff to ensure the awareness of the need to improve (DeuL2)	*	*				*	*

Double-loop learning (DL)

Double-loop learning (DL) refers to a revision of assumptions and actions after undertaking a comprehensive review of the root causes of errors. An example of DL would be when a coach evaluates with his professional road cycling team what went wrong in previous legs of a race, in order to determine a better racing strategy for the next leg. When undergoing DL, organisations distinguish and tackle the root causes of underperformance and improve their ways of working. Jashapara (2003) also noted that organisations determine their new priorities and weighting of organisational norms in the presence of DL when the assumptions and principles that comprise the leading variables are being challenged. Wong *et al.* (2009) described that DL acts as an indicator and initiative for organisations to take corrective actions after revisiting the underlying assumptions. DL focuses on the need to search for alternatives instead of modifications, in order to avoid repetition of the same problems.

The practice of DL can be characterised by: (i) identifying the root of the problem before taking improvement action (DL1); and (ii) Seeking and adopting new management and working approaches through evaluation of current practice (DL2) (see Table 5.1).

Deutero-learning (DeuL)

Deutero-learning (DeuL) refers to the capacity of learning to learn (Argyris and Schon 1978). Francis (1997) described DeuL as a mechanism or system development 'which forces learning to become explicit, and it is the avenue for organisations to leverage a continuing commitment to learning'. In order to accomplish DeuL, organisations should map out all areas that contribute to its improvement as an entity and set out to improve its ability to learn effectively in each of the areas that constitutes its total learning (Pedler *et al.* 1997). Kululanga *et al.* (1999) emphasised the necessity of DeuL style for organisations to attain sustainable performance improvements. While SL and DL types are concerned with 'the operational events which are the subject of learning' (Francis 1997), DeuL 'conceptualises learning as an event in its own right, making the learning process much more conceptual and transferable' (ibid.). The practice of DeuL can be characterised by: (i) collecting and focusing on information that reflects the need to improve (DeuL1); and (ii) systematising the communication channels among staff to ensure the awareness of the need to improve (DeuL2).

Measuring performance improvement (PER)

Depending on the focus of the studies, researchers measured performance improvement (PER) in terms of the construction contracting organisations' project success (Jashapara 2003). Some of them gauged performance by the construction contracting organisations' achievement in terms of meeting the client's requirement (Tam *et al.* 2004; Xiao and Proverbs 2003), as well as the target of profit that can be derived from the project (Yasamis *et al.* 2001; Xiao and Proverbs 2003). They described PER as the minimisation of deviations between actual and predetermined standards (Holt *et al.* 2000).

Some researchers based their study on the work of Drucker (1974) and argued that PER should not only be gauged by the achievement of measurable benefits, but also by the effectiveness that the construction contracting organisations can attain through sustainable performance improvement (Jashapara 2003, Law and Chuah 2004, Wong and Cheung 2005a). In this regard, PER can be identified by the competence of construction contracting organisations to change for meeting the changing project requirements (Jashapara 2003; Wong and Cheung 2005a; Yasamis *et al.* 2001), address forthcoming risks and consequences (Law and Chuah 2004; Wong and Cheung 2005a) and take prompt actions to tackle recurring problems (Yasamis *et al.* 2001; A1-Jibouri 2003; Enno *et al.* 2003; Wong and Cheung 2005a).

In sum, previous studies suggested that PER can be identified by the attributes as shown in Table 5.2.

The conceptual model

Based on the foregoing sections, a conceptual model for this study is shown in Figure 5.3. The model is based on the work of Argyris and Schön (1978) and identifies the practice of OL by three OL types: single-loop learning (SL), double-loop learning (DL) and deutro-learning (DeuL). Furthermore, performance improvement of the construction contracting organisations is measured by the five attributes listed in Table 5.3. The arrows represent the direction of the hypothesised influence (Bollen and Long 1992; Hoyle 1995; Molenaar *et al.* 2000). For example, the practice of SL can be identified by 'Working (and considering corrective actions if required) under a clearly identified project goal' and 'Referring to the firm's past experience to interpret the performance feedback'. Hence, an arrow shown originates from 'SL' to 'SL1' and 'SL2' respectively. Similarly, the rest of the attributes as

Table 5.2 Attributes for identification of performance improvement (PER)

Attributes	Yasamis et al. 2001	Ward et al. 2001	Al-Jibouri 2003	Xiao and Proverbs 2003	Enno et al. 2003	Jashapara 2003	Tam et al. 2003	Law and Chuah 2004	Wong and Cheung 2005a
Level of meeting the client's requirements (**Per1**)	*	*	*				*		*
Extent of attaining the anticipated profit (**Per2**)	*			*					*
Being competent to change in order to meet the changing project requirements (**Per3**)	*					*			*
Being competent to address forthcoming risks and consequences (**Per4**)								*	*
Being competent to take prompt actions to tackle recurring problems (**Per5**)	*		*	*					*

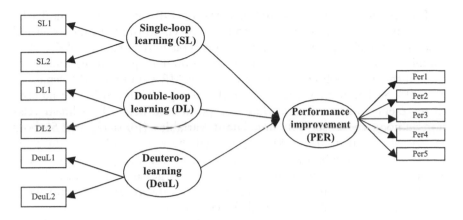

Figure 5.3 The conceptual model for the relationship between OL and continuous improvement

shown in Tables 5.1 and 5.2 are used to measure the practice of DL and DeuL and the continuous improvement of the construction contracting organisations.

Data analysis and methods

In this study, several statistical analyses are considered for testing the hypotheses. To examine the validity of grouping the attributes to the particular OL types, Confirmatory Factor Analysis (CFA) can be used (Sharma 1996). To determine the impact of the OL types implementation on CI, multiple regression analysis (MRA) is available (Norušis 1999). This study employs the two-step analysis approach suggested by Jöreskog and Sörbom (1996) to test model validity. This first step involves the checking of construct reliability and inter-relationships. The checking of the construct reliability aims to validate the internal consistency of the construct (i.e. the reliability of a latent variable to be presented by its observed variables). This can be done by conducting Cronbach alphas reliability testing. The alpha value ranges from 0 to 1. The higher the alpha value, the greater the internal consistency of the construct is. A value from 0.6 to 0.7 is regarded as 'sufficient' and a value higher than 0.7 is regarded as 'good' (Sharma 1996). The checking of the construct inter-relationships, which can be done by conducting Pearson correlation analysis, aims to validate the proposed inter-relationships among constructs (i.e. the convergent validity of the conceptual model). The applications of the Cronbach alphas reliability testing and the Pearson correlation analysis within this study were executed by using the Statistical Package for Social Sciences (SPSS) – Version 11.0.

The second step involves analysing the overall fitness of the model by investigating the fitness of the hypothesised relationships using structural equation modelling (SEM) analyses (Jöreskog and Sörbom 1996). SEM technique has been used in several construction management research studies and this is a method that integrates both MRA and CFA (Molenaar *et al.* 2000; Sarkar *et al.* 1998; Wong and Cheung 2005b). SEM can be used to represent, estimate and validate a hypothesised network of linear relations among the observable and latent variables (Jöreskog and Sörbom 1996). Molenaar *et al.* (2000) emphasised the use of SEM to reduce the shortcomings brought by the MRAs because the SEM framework also account for the errors in measurement when a large number of variables are involved. Thus, a more holistic representation of the intertwined framework can be evaluated produced through SEM analysis.

A computer package called Analysis of Moment Structures 5.0 (AMOS) is used for the SEM analysis. The validity of the conceptual model is assessed in terms of the Goodness of Fit (GOF) indices. There are several GOF indices available from AMOS to test the conceptual model validity. The recommended levels of the GOF indices are shown in Table 5.3. If the conceptual model could not reach the recommended levels, model refinements are required. AMOS offers a modification indices function that gives recommendations on how to improve the GOF values. The recommendations include revising the interrelationship paths and adding covariance error paths between observed and latent variables. The modification actions should not be taken intuitively and incidentally (Arbukle and Wothke 1999). They 'must only be considered if they make theoretical sense' (Arbukle and Wothke 1999).

Data collection

Data was collected by way of a questionnaire derived from the conceptual model. The questionnaire has two parts. The first part includes questions designed to solicit the respondents' demographic information. In the second part, respondents were asked to assess their company's practice of the OL types in a construction project and the level of project performance improvements. The data collection questionnaire is given in Appendix 5A.

In this study, the targeted respondents were randomly selected from the consultants, contractors and suppliers firms or the public sector departments listed in the latest edition of the Hong Kong Builders' Directory (Far East Trade 2003), and the official webpage of the professional institutes such as the Hong Kong Institute of Architects (HKIA), Hong Kong Institution of Engineers (HKIE) and the Hong Kong Institute of Surveyors (HKIS). They are directors, project managers and professional grade staffs (including engineers and surveyors). A total of 300 questionnaires were sent to the identified respondents by post or fax. Ninety-one questionnaires were returned. Eight of them were incomplete and thus discarded from the data

Table 5.3 Recommended levels of the GOF measures (adapted from Molenaar *et al.* 2000)

Goodness-of-fit (GOF) measure	Description	Recommended level of GOF measure
X^2 / degree of freedom	Chi-square/df ratio is the minimum discrepancy divided by its degrees of freedom. Several writers have suggested that the use of this ratio as a measure of fit. For every estimation criterion, the ratio should be close to one for correct models. The ratio is indicative of an acceptable fit in the ranges of 2 to 1 or 3 to 1 between the hypothetical model and the sample data (Carmines and McIver 1981).	Recommended level from 1 to 2
Goodness-of-fit index (GFI)	The purpose of computing GFI is to maximise the likelihood estimate. GFI is always between 0 and 1, where unity indicates a perfect fit (Arbukle and Wothke 1999).	0 (No fit) to 1 (Perfect fit)
RMSEA	Practical experience has made us feel that a value of the RMSEA of about 0.05 or less would indicate a close fit of the model in relation to the degrees of freedom. This figure is based on subjective judgment. It cannot be regarded as infallible or correct, but it is more reasonable that the requirement of exact fit with the RMSEA = 0.0. Browne and Cudeck (1993) suggest that a RMSEA of .05 or less indicates a 'close fit'.	<0.05 indicates very good fit – threshold level is 0.10
p-close (at RMSEA<0.05)	P-close value indicates the probability of getting a 'close fit' model (i.e. RMSEA <0.05) (Browne and Cudeck 1993). The typical range for P-close lies between 0 and 1. The higher the P-close value indicates the greater probability of getting a 'close fit' model (i.e. RMSEA <0.05) by the use of the sample (Arbukle and Wothke 1999).	0 (No fit) to 1 (Perfect fit)
Tucker-Lewis index (TLI)	The Tucker-Lewis coefficient was discussed by Bentler and Bonett (1980) in the context of analysis of moment structures, and is also known as the Bentler–Bonett non-normed fit index (NNFI). The typical range for TLI lies between 0 and 1, but it is not limited to that range. TLI values close to 1 indicate a very good fit.	0 (No fit) to 1 (Perfect fit)
Normal fit index (NFI)	Normal fit index (NFI) was discussed by Bentler and Bonett (1980) in the context of analysis of moment structures. The typical range for NFI lies between 0 and 1, but it is not limited to that range and NFI values close to 1 indicate a very good fit.	0 (No fit) to 1 (Perfect fit)

Table 5.4 Questionnaires sent and received

	Developers (public and private)	Consultants	Contractors and suppliers	Total
Sent (No.)	50	100	150	300
Received (No.)	15	21	47	83
% Received	30	21	31.3	27.7

analyses. As a result, a total of 83 usable responses were used in the analysis (see Table 5.4). The valid response rate, therefore, is 27.7 per cent. Easterby-Smith *et al.* (1991) opined that the reasonable response rate of questionnaire survey studies conducted in the construction industry ranges from 25 per cent to 30 per cent. Similar research studies about OL in construction conducted by Kululanga *et al.* (1999) were based on 31 responses (equivalent to 34 per cent of the response rate). The response rate of the research study about the effect of OL on contractors' performance conducted by Jashapara (2003) was 14.1 per cent. Therefore, both sample size and return rate in this study are considered reasonable.

It is also noted that a sample size of 100 (preferably 200) is recommended for use in SEM analyses (Hair *et al.* 1998; Jöreskog and Sörbom 1998; Kline 1998). In addition, Hair *et al.* recommended that the ratio between the sample size and the number of free parameters to be 5:1 under normal distribution theory. Otherwise, the estimated regression weights of both latent and independent variables may become statistically insignificant with high standard errors. Notwithstanding, Hair *et al.* suggested that even a small sample size of 50 may provide valid results in SEM analyses. For example, Ozorhon *et al.* (2008) employed SEM analysis with 67 data sets to investigate the implications of culture in the performance of the international construction joint venture projects. The SEM-based studies by Petersen *et al.* (2000), Smith and Smith (2004) and Paiva *et al.* (2008) also involved fewer than 100 data sets.

Analyses and results

Step 1: checking construct reliabilities and inter-relationships

In order to ensure the appropriateness of groupings of the variables to the respective constructs as shown in the conceptual model (see Figure 5.3), internal consistency of the constructs has to be checked. Table 5.5 details the results obtained from the Cronbach alphas reliability testing. All groupings have the Cronbach alpha values above 0.7, which suggest that the variables are significantly related to their specified constructs (Hair *et al.* 1998). Hence, all variables included in the conceptual model, as well as their respective groupings are retained in the initial SEM (Jashapara 2003).

Table 5.5 Descriptive statistics, internal consistency and correlations

| Construct | Alpha | Mean | S.D. | Pearson correlation | | | |
				SL	DL	DeuL	PER
Single-loop learning (SL)	0.70	4.72	.75	1			
Double-loop learning (DL)	0.70	4.67	.56	.78(**)	1		
Deutero-learning (DeuL)	0.75	4.81	.60	.71(**)	.80(**)	1	
Performance improvement (Per)	0.81	4.56	.51	.51(**)	.44(**)	.31(**)	1

Notes: Alpha is the standardised Cronbach alpha coefficient; ** Correlation is significant at the 0.01 level (2-tailed)

Pearson correlation analyses were used to validate the relationships between the practice of the OL types and PER as proposed in the conceptual model (Jöreskog and Sörbom 1996; Jashapara 2003). The results are shown in Table 5.5. Significant relationships (at p < 0.01 level) between the practice of OL types and PER supported the convergent validity of the conceptual model (Jashapara 2003).

Step 2: SEM analysis

The initial model

The specifications of the initial structural model follow the conceptual model as shown in Figure 5.3. The GOF results of the initial model are also shown in Table 5.6, the indices marked with '*' are items that do not reach the recommended levels. As such, model refinements are needed until all GOF measures pass the recommended levels (Wong and Cheung 2005b).

Table 5.6 Goodness-of-fit (GOF) of the initial and final models (adapted from Molenaar *et al.* 2000)

| Goodness-of-fit (GOF) measure | Recommended level of GOF measure | Model's GOF | |
		Initial	Final
X² / degree of freedom	Recommended level from 1 to 2	3.47*	1.02
Goodness-of-fit index (GFI)	0 (No fit) to 1 (Perfect fit)	0.74	0.93
RMSEA	<0.05 indicates very good fit – threshold level is 0.10	0.17*	0.02
p-close	p-value for hypothesis testing of RMSEA	0.00	0.74
Tucker-Lewis index (TLI)	0 (No fit) to 1 (Perfect fit)	0.56	0.99
Normal fit index (NFI)	0 (No fit) to 1 (Perfect fit)	0.60	0.90

The final model

Model refinements can be achieved by revising the interrelationship paths and adding covariance error paths between variables and the latent factors (Arbukle and Wothke 1999). However, the revision of the interrelationship paths as well as the addition of the covariance error paths should not be conducted intuitively or arbitrarily. Any revision to be made must have a theoretical base (ibid.). With reference to the suggestions from the modification indices available from AMOS and the theoretical implications of the model refinements, a refined model (final model) is developed and shown in Figure 5.4.

Discussion

It is found that the final model passed all the GOF requirements (see Table 5.6). The ratio for $\chi2/df$ was 1.02 and the Goodness of Fit index value (GFI) was 0.93, both indices indicate that the final model provides a good fit to the data. The Root Mean Square Error of Approximation (RMSEA) value was 0.02 at $p<0.05$. As such, a RMSEA value which is lower than 0.10 represents the significance of the final model (Wong and Cheung 2005b).

When the final model was compared with the initial SEM, no attribute and latent variable was eliminated. Moreover, in the final model, all path coefficients were positive and all path relationships were significant at $p <0.05$. Nevertheless, some of the interrelationship paths were revised (see Figure 5.4). It is noted that the relationship path between DeuL and PER were delinked. Instead DeuL is now linked with SL and DL. DeuL has been

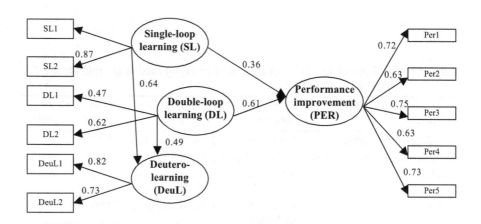

Figure 5.4 The final model with standardised regression weights
Note: All path coefficients are significant at $p <0.05$

described as the use of systems to facilitate continuous learning of project participants (Pedler *et al.* 1997; Redding and Catalanello 1994). In this respect, DeuL may not work as a learning type per se, instead it represents the systems and platforms that facilitate the achievement of SL and DL. It makes good sense as the DeuL attributes are indeed system-based or strategy-oriented when compared to the more operational nature of the attributes of SL and DL (Tosey 2005). In fact, Kululanga *et al.* (1999) described DeuL as 'learning to learn' and is a core competence for a construction contracting organisation that strives for continuing improvement. Furthermore, the path coefficient between DeuL and SL is 0.64 as compared to 0.49 for DeuL and DL. This may be interpreted as SL is more responsive to DeuL arrangements.

Turning now to the relationship links between learning styles and project performance, the final SEM model (Figure 5.4) suggests greater impact on performance by practicing DL when compared with SL. With a quick glance of the learning attributes, it can be noted that SL attributes are more directed for front-line operations and drive for 'quick fix' whereas DL practices call for the re-examination of some, if not all, of the fundamentals. The changes derived therefrom would probably be more radical, yet more enduring and having greater impact. Jashapara (2003) shared similar insight as his study suggested that project performance is mainly affected by DL rather than SL practices.

It is surprising to note the path coefficient between SL and PER is only 0.36. Perhaps the collective wisdom of the respondents is that SL practices only provide symptomatic treatments and would hardly bring forth perpetual efficiency and effectiveness enhancement (Kululanga *et al.* 1999). This explanation may perhaps be too intuitive and investigation in this connection is thus suggested for further research.

This study examines the effect of different learning types on project performance. The findings of this study are beneficial to construction contracting organisations by highlighting the importance of enabling a learning organisation. The empirical testing of the conceptual model confirms this. In addition, the ways and means to attain learning are also made available by examining the learning attributes of SL, DL and DeuL.

Summary

The ability to remain competitive is itself a power of an organisation. Learning organisation has emerged as one of the key ways to allow organisations to secure efficiency and effectiveness. Construction activities, in particular, are prone to error due to the massive on-site operations. Coincidently, these errors are very costly, for the cost of rectification, disruption and delay. The ability to learn and improve from mistakes should therefore be a core competence of every construction contracting organisation.

A study was carried out in Hong Kong to conceptualise and test empirically the relationship between learning styles and project performance improvement. Previous studies identified three major learning styles: single-loop learning (SL), double-loop learning (DL) and deutero-learning (DeuL). It is hypothesised that all three types of learning styles contribute to project performance improvement measured in terms of efficiency and effectiveness. To test the hypothesis, the technique of structural equation modelling was employed. Measures of SL, DL, DeuL, project efficiency and project effectiveness were collected through a questionnaire survey. Attribute statements developed from previous studies were used as measures.

Four key findings are noted. First, DeuL is not directly linked with project performance improvement. Instead, it serves as the platform to facilitate the achievement of SL and DL. Second and in this connection, SL appears to be more responsive to DeuL arrangement as compared to DL. This is reflected by the higher path coefficient values of the final SEM model. Third, DL was found having greater impact on project performance. This is probably the result of DL practices aim to re-examine the underlying assumption, whereas SL practices are mainly directed for 'quick-fix'. Last, negative path coefficient value was recorded between practicing SL and project performance improvement. This may be the 'quick fix' nature of SL practice only providing symptomatic treatments and thus is not conducive for perpetual efficiency and effectiveness enhancement.

Acknowledgements

Special thanks to Ms. Ka Lam Fan for collecting data for the study. The content of this chapter has been published in Volume 135(6) of the *Journal of Construction Engineering and Management* and is used with the permission from ASCE.

References

Al-JiBouri, S. H. (2003) Monitoring systems and their effectiveness for project cost control in construction. *International Journal of Project Management*, 21(3), 145–54.

Arbukle, J. L. and Wothke, W. (1999) *Amos 4.0 User's Guide*. USA: Small Waters Corporation.

Argyris, C. and Schön, D. (1978) *Organisational Learning: A Theory of Action Perspective*. Reading, MA: Addison-Wesley.

Baltrusch, R. (2001) Exploring organisational learning in virtual forms of organization, in *Proceedings of the 34th Hawaii International Conference on System Sciences*.

Bennis, W. and Nanus, B. (1985) *Strategy IV: Deployment of Self, Leaders: Strategies for Taking Charge*. New York: Harper & Row.

Bentler, P. M. and Bonett, D. G. (1980) Significant tests and goodness of fit in the analysis of covariance structures. *Psychological Bulletin*, 88, 588–606.

Bollen, K. A. and Long, J. S. (1992) Tests for structural equation models. *Special issue of Sociological Methods & Research*, 21, 123–282.

Bresnen, M., Edelman, L., Newell, S., Scarbrough, H. and Swan, J. (2003) Social practices and the management of knowledge in project environments. *International Journal of Project Management*, 21(3), 157–66.

Bröchner, J., Adolfsson, P. and Johansson, M. (2002) Outsourcing facilities management in the process industry: a comparison of Swedish and UK patterns. *Journal of Facilities Management*, 1(3), 265–71

Carmines, E. and McIver, J. (1981) Analyzing models with unobserved variable, in G. Bohrnstedt and E. Borgatta (eds), *Social Measurement: Current issues*. Beverly Hills: Sage, 65–115.

Cayes, K. (1998) The need to learn, and why engineers may be poor students. *Journal of Management in Engineering*, 14(2), 31–33.

Chaaya, M. and Jaafari, A. (2001) Cognizance of visual design management in life-cycle project management. *Journal of Management in Engineering*, 17(1), 49–57.

Construction Industry Review Committee of Hong Kong (CIRC) (2001) *Construct for Excellence*. Hong Kong: Construction Industry Review Committee.

Crawford, P. and Bryne, P. (2003) Project monitoring and evaluation: a method for enhancing the efficiency and effectiveness of aid project implementation. *International Journal of Project Management*, 21(5), 363–73.

Davey, C., Powell, J. A., Cooper, I. and Powell, J. E. (2004) Innovation, construction, SMEs and action learning. *Engineering, Construction and Architectural Management*, 11(4), 230–37.

Dorée, A. G. and Holmen, E. (2004) Achieving the unlikely: innovating in the loosely coupled construction system. *Construction Management and Economics*, 22(8), 827–38.

Drucker, P. F. (1974) *Management: Tasks, Responsibilities, Practices*. London: Heinemann.

Duncan, R. and Weiss, A. (1979) Organizational learning: implication for organization design. *Research in Organizational Behaviour*, 1, 75–123.

Easterby-Smith, M., Thorpe, R. and Lowe, A. (1991) *Management Research: An Introduction*. London: Sage Publications.

Egan, J. (1998) *Rethinking Construction: The Report of the Construction Task Force*. London: Department of the Environment, Transport and the Regions.

Enno, E. D., Koehn, P. E. and Datta, N. K. (2003) Quality, environmental, and health and safety management systems for construction engineering. *Journal of Construction Engineering and Management*, 129(9), 562–69.

Far East Trade (2003) *Hong Kong Builder Directory* (no. 35). Hong Kong: Far East Trade Press.

Fiol, C. M. and Lyles, M. A. (1985) Organisational learning. *Academy of Management Review*, 10(4), 80–13.

Ford, D. N., Voyer, J. J. and Gould Wilkinson, J. M. (2000) Building learning organization in engineering cultures: case study. *Journal of Management in Engineering*, 16(4), 72–83.

Francis, S. (1997) A time for reflection: learning about organizational learning. *The Learning Organization*, 4(4), 168–79.

Franco, L. A., Cushman, M. and Rosenhead, J. (2004) Project review and learning in construction industry: embedding a problem structuring method within a partnership context. *European Journal of Operational Research*, 152(6), 586–601.

Garvin, D. (1993) Building a learning organisation. *Harvard Business Review*, July–August, 78–91.

Gherardi, S. (2002) Learning: organizational, in N. J. Smelser and James Wright (eds), *International Encyclopedia of the Social and Behavioral Sciences*. Oxford: Elsevier, 8609–13.

Goulding, J. S. Alshawi, M. (2004) A process-driven IT training model for construction: Core development issues. *Construction Innovation: Information, Process, Management*, 4(4), 243–54.

Hair, J. F., Tatham, R. L., Anderson, R. E. and Black, W. (1998) *Multivariate Data Analysis* (5th edn). London: Prentice-Hall.

Hayes, J. and Allinson, C. W. (1998) Cognitive style and the theory and practice of individual and collective learning in organisations. *Human Relations*, 51, 847–71.

Holt, G. D., Proverbs, D. and Love, P. E. D. (2000) Survey findings on UK construction procurement: is it achieving lowest cost, or value? *Asia Pacific Building and Construction Management Journal*, 5(2), 13–20.

Hoyle, R. H. (ed.) (1995) *Structural Equation Modeling: Concepts, Issues, and Applications*. Thousand Oaks, CA: Sage Publications.

Huber, G. P. (1991) Organizational learning: the contributing processes and the literatures. *Organization Science*, 2(1), Special Issue, 88–115.

Huemer, L. and Östergren, K. (2000) Strategic change and organizational learning in two Swedish construction firms. *Construction Management and Economics*, 18(6), 635–42.

Jashapara, A. (2003) Cognition, culture and competition: an empirical test of the learning organization. *The Learning Organization*, 10(1), 31–50.

Jöreskog, K. and Sörbom, D. (1996) *LISREL 8: Users Reference Guide*. Hillsdale, NJ: Lawrence Erlbaum Associates.

Kale, S. and Arditi, D. (1999) Age-dependent business failures in the US construction industry. *Construction Management and Economics*, 17(4), 493–503.

Kline, R. B. (1998) *Principles and Practices of Structural Equation Modeling*. New York: Guilford.

Kululanga, G. K. and McCaffer, R. (2001) Measuring knowledge management for construction organizations. *Engineering, Construction and Architectural Management*, 8, 346–54.

Kululanga, G. K., McCaffer, R., Price, A. D. F. and Edum-Fotwe F. (1999) Learning mechanisms employed by construction contractors. *Journal of Construction Engineering and Management*, 125(4), 215–33.

Kululanga, G. K., Edum-Fotwe, F. T. and McCaffer, R. (2001) Measuring construction contractors' organisational learning. *Building Research and Information*, 29(1), 21–29.

Kululanga, G. K., Price, A. D. F., McCaffer, R. (2002) Empirical investigation of construction contractors' organizational learning. *Journal of Construction Engineering and Management*, 128(5), 385–91.

Kumaraswamy, M. M. (1998) Industry development through creative project packaging and integrated management. *Engineering, Construction and Architectural Management*, 5(3), 228–37.

Kurtyka, J. (2003) Implementing business intelligence systems: an organizational learning approach. *DM Review Magazine*, November, 2003. Available at www.dmereview.com/editorial/dmreview/print_action.cfm?articleId=7610

Law, K. M. Y. and Chuah, K. B. (2004) Project-based action learning as learning approach in learning organization: the theory and framework. *Team Performance Management*, 10, 7/8, 178–86.

Love, P. E. D. and Josephson, P.-E. (2004) Role of error-recovery process in projects. *Journal of Management in Engineering*, 20(2), 70–79.

Love, P. E. D., Huang, J. C., Edwards, D. J. and Irani, Z. (2004) Nurturing a learning organization in construction: a focus on strategic shift, organizational transformation, customer orientation and quality centred learning. *Construction Innovation*, 4(2), 113–26.

Love, P. E. D., Li, H., Irani, Z. and Faniran, O. (2000) Total quality management and the learning organization: a dialogue for change in construction. *Construction Management and Economics*, 18(3), 321–31.

Mintzberg, H. (1989) *Mintzberg on Management: Inside our Strange World of Organizations*. New York: The Free Press.

Molenaar, K., Washington, S. and Diekmann, J. (2000) Structural equation model of construction contract dispute potential. *Journal of Construction Engineering and Management*, 126(4), 268–77.

Murray, P. and Chapman, R. (2003) From continuous improvement to organizational learning: developmental theory. *The Learning Organization*, 10(5), 272–82.

Norušis, M. J. (1999) *SPSS 9.0 Guide to Data Analysis*. Englewood Cliffs, NJ: Prentice Hall.

Oyegoke, A. S. (2006) Building competence to manage contractual claims in international construction environment: the case of Finnish contractors. *Engineering, Construction and Architectural Management*, 13(1), 96–113.

Ozorhon, B., Dikmen, I., and Birgonul, M.T. (2005) Organizational memory formation and its use in construction. *Building Research and Information*, 33(1), 67–79.

Ozorhon, B., Arditi, D., Dikmen, I. and Birgonul, M.T. (2008) Implications of culture in the performance of international construction joint venture. *Journal of Construction Engineering and Management*, 134(5), 361–70.

Paiva, E. L., Roth, A. V. and Fensterseifer, J. E. (2008) Organizational knowledge and manufacturing strategy process: a resource-based review analysis. *Journal of Operational Management*, 26(1) 115–32.

Pawlowsky, P. (2001) Management science and organizational learning, in M. Dierkes *et al.* (eds), *The Handbook of Organizational Learning and Knowledge*. Oxford: Oxford University Press, 61–88.

Pedler, M., Burgoyne, J. and Boydell, T. (1997) *The Learning Company: Strategies for Sustainable Development*. London: McGraw-Hill.

Petersen, K. J., Frayer, D. J. and Scannell, T. V. (2000) An empirical investigation of global sourcing strategy effectiveness. *Journal of Supply Chain Management*, 36(2), 29–38.

Redding, J. C., and Catalanello, R. F. (1994) *Strategic Readiness: The Making of Learning Organizations*. San Francisco, CA: Jossey-Bass.

Sarker, M. B., Aulakh, P. S. and Cavusgil, S. T. (1998) The strategic role of relational bonding in inter-organizational collaborations: an empirical study of the global construction industry. *Journal of International Management*, 4(2), 415–21.

Sharma, S. (1996) *Applied Multivariate Techniques*. New York: John Wiley & Sons.

Smith, D. and Smith, K. L. (2004) Structural equation modeling in management accounting research: Critical analysis and opportunities. *Journal of Accounting Literature*, 23, 49–86.

Tam, C. M., Tam, V. W. Y. and Tsui, W. S. (2004) Green construction assessment for environmental management in the construction industry of Hong Kong. *International Journal of Project Management*, 22, 563–71.

Tosey, P. (2005) The hunting of the learning organization: a paradoxical journey. *Management Learning*, 36(3), 335–52.

Visser, M. (2004) Deutero-learning in organizations: A review and a reformulation, *Working paper series on research in relationship management,* RRM-2004-07-MGT, Nijmegen School of Management, University of Nijmegen. Available from: www.nsm.kun.nl

Wong, P. S. P. and Cheung, S. O. (2005a) From monitoring to learning: a conceptual framework, in *Proceedings of the 21st Annual Conference of the Association of Researchers in Construction Management*, SOAS, London, 1037–51.

Wong, P. S. P. and Cheung, S. O. (2005b) A structural equation model on trust and partnering success. *Journal of Management in Engineering*, 21(2), 70–80.

Wong, P. S. P. and Cheung, S. O. (2008) An analysis of the relationship between learning behaviour and performance improvement of the contracting organizations. *The International Journal of Project Management*, 26(2), 112–23.

Wong, P. S. P., Cheung, S. O. and Fan, J. K. L. (2009) Examining the relationship between organizational learning styles and project performance: a structural equation modeling approach. *Journal of Construction Engineering and Management*, 135(6), 497–507.

Xiao, H. and Proverbs, D. (2003) Factors influencing contractor performance: an international investigation. *Engineering, Construction and Architectural Management*, 10(5), 322–32.

Yasamis, F. Arditi, D. and Mohammadi, J. (2001) Assessing contractor quality performance. *Construction Management and Economics*, 20(3), 211–23.

Yashiro, T. (2001) A Japanese perspective on the decline of robust technologies and changing technological paradigms in housing construction: issues for construction management research. *Construction Management & Economics*, 19(3), 301–6.

Zipf, P. J. (1998) An integrated project management system. *Journal of Management in Engineering*, 14(1), 38–41.

Appendix 5A

Part 1: Personal Information:

Q1.1 Type of your company: a: Developer   b: Contractor
                            c: Consultant   d: Others: _____

Q1.2 Working experience in   a: <5 years   b: 5–10 year
      the construction field: c: 11–15 years   d: 16–20 years
                               e: >20 years

With reference to one construction **project** that you have been (either fully or partly: involved for at least 1 year and provide the following particulars:

Q1.3 Project   1) Residential   2) Office   3) Hotel
      nature   4) Shopping   5) Infrastructure   6) Complex:
                    centre and/or   comprising
                    carpark   1) and 4)

                 7) Complex: 8) Complex: 9) Others:
                    comprising   comprising
                    2) and 4)   3) and 4   _____

Q1.4 Project name: _____

Part 2: Measure of organizational learning

Do you agree that your firm practiced the followings during the project

Q2.1 Working (and considering corrective actions   1 2 3 4 5 6 7
      if required) under a set of clearly identified
      project goals.

Q2.2 Referring the firm's past experience to interpret 1 2 3 4 5 6 7
      the performance feedback.

Q2.3 Identifying the root of the problem before   1 2 3 4 5 6 7
      taking improvement action.

Q2.4 Seeking and adopting new management and   1 2 3 4 5 6 7
      working approach through evaluation of
      current practice.

Part 3: Measure of performance improvement

Do you agree with the following statements:

Q3.1 Your firm's service quality in delivering the   1 2 3 4 5 6 7
      project met or exceeded the client's requirements.

Q3.2 Your firm met or exceeded the anticipated profit   1 2 3 4 5 6 7
      through delivering this project.

Q3.3 Your firm is competent to change in order to   1 2 3 4 5 6 7
      meet the changing project requirements.

Q3.4 Your firm is competent to address forthcoming   1 2 3 4 5 6 7
      risks and consequences.

Q3.5 Your firm is competent to take prompt actions   1 2 3 4 5 6 7
      to tackle the recurring problems

6 Exploring the traction between intra-organisational and inter-organisational learning

Peter Shek Pui Wong

Introduction

Learning is vital for contracting organisations to sustain continuous improvement and competitive advantages (see Table 6.1). In the last chapter, it has been reported that practising single-loop learning (SL) and double-loop learning (DL) was found to have a significant positive effect on performance improvement. Furthermore, construction contracting organisations' practice of deutero-learning (DeuL) was found to be imperative for their practice of the SL and DL. The reported study focused at how a single contracting organisation practiced learning in a project environment.

However, in view of the fact that learning can occur within and among contracting organisations, researchers advocated that OL in construction should also be studied at an inter-organisational level (Chan *et al.* 2005). To this end, researchers conceptualised performance improvement as not merely an outcome of learning by a single contracting organisation, but among the collaborating firms in a construction project. They further distinguished OL in two forms: intra-organisational learning (intra-OL) and inter-organisational learning (inter-OL) (Bapuji and Crossan 2004; Chan *et al.* 2005).

Intra-OL has been described intrinsically and extrinsically (Chan *et al.* 2005; Holmqvist 2003; Bapuji and Crossan 2004). Intrinsically, intra-OL has been described as a mechanistic stimulus-response process (Couto and Teixeira 2005; Kagioglou *et al.* 2000; Pawlowsky 2001). It has been assumed that performance improvement can be attained through repetitive tasks (Cyert and March 1963; Pawlowsky 2001). Extrinsically, intra-OL has been described as a learning process that is triggered by the change of business environment (Chan *et al.* 2005). Contracting organisations are assumed to learn by adapting to the changes in market demand for performance improvement (Argyris and Schön 1978; Jashapara 2003; Murray and Chapman 2003). The practice of intra-OL can be identified by the contracting organisations' practices of SL, DL and DeutroL.

On the other hand, inter-OL has been described as a process of experience sharing and communication among contracting organisations for attaining performance improvement (Chan *et al.* 2005). Based on a review

Table 6.1 The importance of learning for contracting organisations in construction

Learning is vital for contracting organisations to	Descriptions
Sustain continuous improvement	Learning offers avenues for contracting organisations to bring about a continuous improvement agenda in the construction operational process (Kululanga *et al.* 1999).
	Organisation's inability to learn is a precursor of the project performance improvement. Therefore, organisations should embrace learning in construction projects (Ford *et al.* 2000).
	Contracting organisations' performance has been declining in terms of the erosion of productivity, reduction in profitability and mounting inter-firm adversarial relationships within the supply chain. In order to rectify the problems, exploration of the possible means to facilitate organisation's learning in project-based environment is required (Tjandra and Tan 2002).
	Learning is a source of continuous performance improvement for organisations (Jashapara 2003).
Sustain competitive advantages in a turbulent construction market	Forming learning alliances between firms in a construction project is essential in perspective of sustaining their competitive advantages in the industry (Holt *et al.* 2000).
	Business environment in construction is no longer static. In response to the changing market demands, contracting organisations should equip themselves to become a competent learner (Murray and Chapman 2003).
	Learning is a strategy for contracting organisations to survive in a turbulent construction market (Chan *et al.* 2005).
	Learning is a key for sustaining the organisation's competitive advantages (Siriwardena and Kagioglou 2005).

of literatures, Styhre *et al.* (2004) defined inter-OL as a strategy that fosters experience sharing and communication among contracting organisations in a construction project. Holt *et al.* (2000) highlighted that in a turbulent and changing market which demands focus and flexibility, contracting organisations should learn from each other. They described strategic alliance as a facilitator of inter-OL as it helps creating a 'reflective and mutual learning environment' among organisations and enables them to 'benefit from shared knowledge'.

An assortment of studies about intra-OL was reported in recent years (Chan *et al.* 2005; Love *et al.* 2004; Siriwardena and Kagioglou 2005; Wong and Cheung 2005). In particular, the effect of intra-OL on performance improvement has attracted much research attention (Bresnen and Marshall 2000; Couto and Teixeria 2005; Farghal and Everett 1997; Fu *et al.* 2000; Love and Josephson 2004). By fitting the historical performance data of contractors to learning curves, Farghal and Everett (1997) demonstrated that there is a link between experience accumulation and performance improvement. By analysing the tender information of 266 building contracts in Hong Kong, Fu *et al.* (2000) found that the contractors' bidding competitiveness is positively correlated to their frequency of bidding. Based on case studies on construction projects in Sweden, Love and Joesphson (2004) found that greater project cost savings are attained by those contractors who can effectively learn from experience.

Nevertheless, Chan *et al.* (2005) commented that these studies mainly focused on the learning behaviour at intra-organisational level. From a sociological perspective, they argued that the learning behaviour of the contracting organisations is somewhat related to their inter-OL learning activities. In this connection, they advocated that the effect of inter-OL on the practice of intra-OL should be investigated. Likewise, Love *et al.* (2004) argued that contracting organisations' practice of intra-OL is affected by sociological factors such as the transient nature of construction project teams. As the temporary nature of the project team offers no guarantee of future contracts among team members, contracting organisations may lack the degree of proximity to develop learning from each other. Based on a literature review, Ruuska and Vartiainen (2005) suggest that performance improvement in construction projects is related to the contracting organisations' learning from performance feedback derived from the project monitoring system (PMS). Nevertheless, they argued that some PMS are not well designed in providing useful feedback, in particular to facilitate learning from mistakes. Based on several case studies, Li *et al.* (2001) reported that team-building activities facilitate inter-OL. Furthermore, contracting organisations engaged in inter-OL performed better than those without such engagement. Similar findings were also reported by Holmqvist (2003) who pinpointed that engagement in inter-OL can in fact energise the practice of intra-OL. In sum, previous research, mainly based on literature reviews and case studies, supports the link between inter-OL and intra-OL. Nonetheless, the effect of the engagement in inter-OL on the practice of intra-OL has yet to be systematically investigated (Chan *et al.* 2005).

The concept of inter-organisational learning in construction

Researchers highlighted that inter-organisational learning (inter-OL) is taking place at a project level with temporary coalitions of various organisations (Cheng and Li 2001; El-Diraby and Zhang 2006; Huemer and Östergren 2000; Li *et al.* 2001; Styhre *et al.* 2004). They conceptualised inter-OL as a process of producing sets of inter-organisational strategies for contracting organisations to acquire and integrate knowledge from each other for project performance improvement. Figure 6.1 depicts how previous studies conceptualised intra-OL in construction.

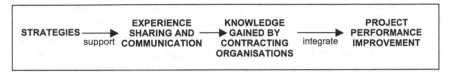

Figure 6.1 Inter-organisational learning process

Generally, inter-OL is described as a learning process taking place among contracting organisations (Chan *et al.* 2005). Chan *et al.* cited the work of Groák (1994) to pinpoint that researchers should abandon the perception of 'one technology, one industry'. They acknowledged that contracting organisations in a construction project may use fundamentally distinct resource and skill bases. For inter-OL to occur, Chan *et al.* suggested the need for strategies that regulate and sustain the experience-sharing and communication between contracting organisations in a project. They used the term 'Network-based OL capabilities' to describe the capability of contracting organisations in terms of their attitudes and successfulness on adopting strategies to facilitate experience-sharing and communication.

Holt *et al.* (2000) highlighted that in a turbulent and changing market that demands focus and flexibility, contracting organisations learn collectively from others. For OL to occur at an inter-organisational level, organisations in a construction project are advised to form strategic alliances as they help create a 'reflective and mutual learning environment' among organisations and enable them to 'benefit from shared knowledge' (Holt *et al.* 2000). Kasvi *et al.* (2003) revealed that facilitating inter-organisational learning requires new knowledge management systems and tools that support these practices.

The question of how inter-OL is practised has drawn much research attention (Chan *et al.* 2005; Ofori 2002). Some studies, based on the literature reviews, described the effect of inter-OL on extrinsic intra-OL of contracting organisations (Huemer and Östergren 2000; Kasvi *et al.* 2003; Ruuska and Vartiainen 2005). Ruuska and Vartiainen (2005) reported that despite the fact that organisations may be suitable for supporting intrinsic intra-OL, they have the tendency to obstruct knowledge sharing and learning among contracting organisations. Kasvi *et al.* (2003) described effective knowledge management as one of the main challenges of project management on account of the varying content and quality of the knowledge and the organisations' low capability to utilise such knowledge. Based on 22 interviews with the staff from two major Swedish contractors' firms, Huemer and Östergren (2000) found that although the firms had been operating successfully in several international projects, they failed to capitalise on the experience learnt in different localities. In other words, no sharing to foster synergy was effected. Styhre *et al.* (2004) studied six construction projects in Sweden and found that OL in construction projects 'does not rely on technical and formal system but rather on personal contacts, communities in practice and learning by doing'. They call for the need to manage knowledge sharing among firms in future. As such, these studies reveal the effect of inter-OL on extrinsic intra-OL. Nevertheless, such effect was rarely backed up with quantitative analysis in previous studies (Chan *et al.* 2005; Ofori 2002).

Hypothesis development

In this chapter, we seek to examine the link between intra-OL and inter-OL, as well as how such relationship may affect the performance of the contracting organisations. To accomplish the research objective, the following hypotheses are tested in this study:

H1: *The practice of intra-OL is contingent on the contracting organisations' engagement in inter-OL.*

H2: *The interactions between the practice of intra-OL and the engagement in inter-OL have moderating effect on performance of the contracting organisations.*

Conceptual model development

The conceptual model is developed by adding a new construct 'Engagement in inter-OL (inter-OL)' to the relationship between intra-OL and performance improvement as suggested in the last chapter. In this connection, attributes for identifying the practice of intra-OL and performance improvement have already been identified and listed in Chapter 5. For ease of reference, they are listed again in Table 6.2.

Construction researchers have used a number of attributes in evaluating contracting organisations' engagement in inter-OL (Chan *et al.* 2005). Kululanga *et al.* (2002) evaluated contracting organisations' engagement in inter-OL by their commitment to seek advice, suggestions or performance improvement methods from others. Furthermore, they also identified that the engagement of inter-OL can be evaluated by the presence of reward schemes; the reward schemes may motivate contracting organisations to learn from others (Kululanga *et al.* 1999; Pedler *et al.* 1997; Redding and Catalanello 1994). Holt *et al.* (2000) included contracting organisations' capability to address and revise the way of working in response to the change of project requirements as a criterion for performance evaluation. Kasvi *et al.* (2003) highlighted that the engagement of inter-OL can be triggered by the development of project monitoring systems (PMS). They described a PMS as a system that records information reflecting project

Table 6.2 Attributes for identifying the practice of intra-OL and performance improvement

Construct	Attributes
Single-loop learning (SL)	Working (and considering corrective actions if required) under a clearly identified project goal (L_1)
	Referring to the firm's past experience to interpret the performance feedback (L_2)
Double-loop learning (DL)	Identifying the root of the problem before taking improvement action (L_3)
	Seeking and adopting new management and working approaches through evaluation of current practice (L_4)
Deutero-learning (DeuL)	Collecting and focusing on information which reflects the need to improve (L_5)
	Systematising the communication channels among staff to ensure the awareness of the need to improve (L_6)
Performance improvement	Level of meeting the client's requirements (P_1)
	Extent of attaining the anticipated profit (P_2)
	Being competent to change in order to meet the changing project requirements (P_3)
	Being competent to address forthcoming risks and consequences (P_4)
	Being competent to take prompt actions to tackle recurring problems (P_5)

performance. They advocated that performance records can be a source of feedback from which lessons can be learnt. In this connection, they conceived that the development of project monitoring systems indicates the intention of contracting organisations to learn in response to feedback. Kasvi *et al.* further stressed that to foster inter-OL, contracting organisations should embrace experience and information sharing under an open and mutual trusting environment. In sum, the engagement in inter-OL can be evaluated by the attributes as shown in Table 6.3.

Based on the hypotheses and the attributes as identified in Tables 6.2 and 6.3, a conceptual model is developed and shown in Figure 6.2. The arrows represent the direction of the hypothesised influence. The conceptual model is underpinned by the work of Argyris and Schön (1978). The practice of intra-OL is identified by the practice of the three OL types: single-loop learning, double-loop learning and deutro-learning (i.e. L1 to L6 as summarised in Table 6.1). The engagement in inter-OL is identified by 8 attributes (i.e. E1 to E8) as summarised in Table 6.2. Furthermore, performance improvement of contracting organisations is assessed by the five attributes as mentioned in Table 6.3 (i.e. P1 to P5).

Table 6.3 Attributes for evaluating the engagement in inter-OL

Engagement in inter-OL in terms of:	Redding and Catalanello (1994)	Pedler et al. (1997)	Kululanga et al. (1999)	Holt et al. (2000)	Barlow (2000)	Kululanga et al. (2002)	Kasvi et al. (2003)	Franco et al. (2004)	Love et al. (2004)	Stybre et al. (2004)
1. Seeking improvement methods from others (E_1) Commit to seek methods for performance improvement from other contracting organisations.							*	*		
2. Addressing changes persistently (E_2) Persistently addressing the change of the project requirements.			*			*			*	
3. Revising the way of working (E_3) Commit to revise the way of working in order to satisfy the changing requirements.			*	*		*			*	
4. Seeking advice and suggestions from others (E_4) Commit to seek advice and suggestions from other contracting organisations.							*	*	*	
5. Openness (E_5) Openness in terms of thoughts and actions about information and experience-sharing with other contracting organisations.								*	*	*
6. Revising long term strategy (E_6) Commit to revise the long term strategy if necessary in order to adapt to the change of project requirements.			*	*		*			*	
7. The development of project monitoring system (E_7) Having an effective project monitoring system established among contracting organisations.							*	*	*	*
8. The presence of incentive scheme (E_8) Reward proposals that encourage staffs to learn from other contracting organisations.	*	*	*							

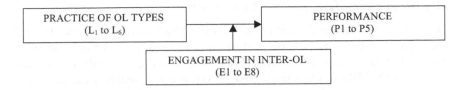

Figure 6.2 Conceptual model about the relationships between engagement of inter-OL, practice of OL types and performance improvement

Research methodology

The questionnaire survey

Data of this study was collected through a questionnaire survey as reported in Chapter 5. The questionnaire consists of 8 questions that are designed for collecting the respondents' perceptive views on their firm's engagement in inter-OL, 6 questions for respondents' evaluation about the practice of intra-OL and 5 questions to assess their firm's performance in a construction project. These questions were evaluated on a seven point Likert scale (from 1: strongly disagree to 7: strongly agree).

Data analysis method

Two data analysis methods were employed in this study: Pearson correlation analysis and multiple moderated regression (MMR) analysis.

Pearson correlation analysis

To examine the relationship between the engagement in inter-OL and practice of intra-OL by contracting organisations (i.e. Hypothesis H1 in this study), Pearson correlation analysis is applied. Pearson correlation analysis is a statistical method with a primary purpose to evaluate whether there is a significant relationship between two sets of ratings. The significance of relationship can be expressed by a ρ-value. When ρ-value is <0.05, the relationship between the two sets of ratings is considered as significant. Furthermore, the extent of the correlation between the two sets of ratings can be signified by the correlation coefficient value (the r-value), which can take on the values from -1.0 to 1.0, where 0 represents no correlation and 1.0 and -1.0 represent perfect positive and negative correlations respectively (Hair *et al.* 1998).

Multiple moderated regression (MMR) analysis

To examine the effect of interaction between the practice of intra-OL and the engagement in inter-OL on performance improvement of the contracting organisations (i.e. Hypothesis H2 in this study), multiple moderated regression (MMR) analysis is applied. MMR is a frequently used technique in management, social and behavioral science research to examine how the relationship between two independent variables affects a dependent variable (Aguinis 1995; Choe 2004; Snell and Dean 1994). Snell and Dean (1994) further described the link between the independent variables as moderating effect. MMR analysis has been applied in previous construction research. Yiu and Cheung (2006) applied MMR to investigate the moderating effect of construction dispute sources on mediation outcomes respective to the tactics used.

The principle of the MMR analysis is founded on multiple regression analysis (MRA) (Cohen *et al.* 2003; Jaccard *et al.* 1990). Aguinis (1995) further described MMR as a two-step approach. Step 1 involves a MRA that assumes that a dependent variable Y can best be predicted by two independent variables X_1 and X_2. The general equation for this MRA can be presented as:

$$Y = a + b_1X_1 + b_2X_2 + \varepsilon \qquad \text{(Eq. 6.1)}$$

Where Y is the dependent variable, X_1, X_2 are the independent variables, a, b_1, b_2 are the unknown constant and ε is the random error for any given set of values for X_1, X_2.

The coefficient of determination, Multiple R-Square (R^2), records the proportion of variation in the dependent variable explained by all the independent variables. The possible value of the measure falls between 0 to 1. The closer R^2 is to 1, the better the correlation provided by the set of independent variables (Hair *et al.* 1998).

Step 2 involves a MRA that includes the moderator in the equation (Cohen *et al.* 2003; Jaccard *et al.* 1990; Yiu and Cheung 2006). Assuming that the predictive power of X_1 on Y varies with the value of X_2, the general equation for the MRA becomes:

$$Y = a + b_1 X_1 + b_2X_2 + b_3X_1X_2 + \varepsilon \qquad \text{(Eq. 6.2)}$$

Cohen and Cohen (1983) named the independent variables X_1 and X_2 as the predictor variables. They further described the product of the predictor variables X_1X_2 as a moderator variable that considers the presence of a moderating effect.

Two criteria can be used to determine whether the moderating effect is significant or not:

1   the inclusion of the moderator variable in Eq. 6.2 produces a significant increase in the squared multiple correlation (R^2) when compared with the simple regression model (Eq. 6.1); and
2   the standardised coefficient of the moderator variable (i.e. b_3) in Eq. 6.2 is significant in explaining the dependent variable (Cohen *et al.* 2003; Cohen and Cohen 1983; Jaccard *et al.* 1990; Yiu and Cheung 2006).

Fisher Z test (F-test hereafter) is used to determine the significance of the ΔR^2 (Yiu and Cheung 2006; Jaccard *et al.* 1990; Cohen *et al.* 2003; Cohen and Cohen 1983). ΔR^2 is considered as significant if the calculated F-value for ΔR^2 is significant at $p < 0.10$ (Enno *et al.* 2003; Hair *et al.* 1998; Law and Chuah 2004). The F-value can be calculated by the following equation:

$$\text{F-value} = [(R_2^2 - R_1^2) / (n_2 - n_1)] / [(1 - R_2^2) / (S - n_2 - 1)] \qquad \text{(Eq. 6.3)}$$

Where
n_1 is the number of predictors in Eq. 6.1;
n_2 is the number of predictors in Eq. 6.2;
S is the total sample size;
$(S - n_2 - 1)$ is the degree of freedom;
R_1^2 is the R^2 value for Eq. 6.1; and
R_2^2 is R^2 value for Eq. 6.2.

T-test for regression coefficient is used to examine the significance of the moderator variable (X_1X_2) in explaining the dependent variable (Y) (Cohen and Cohen 1983). X_1X_2 is considered as significant in explaining Y when the probability of error (p-value) is lower than 0.05 (Cohen and Cohen 1983).

This study employed MMR analyses to examine the effects of engaging different inter-OL attributes on the relationship between practising different OL types and performance improvement. The analyses were done by the Statistical Package for Social Sciences (SPSS) – Version 11.0.

The MMR analyses in this study were developed by the two-step approach proposed by Aguinis (1995). As such, the MMR analyses for this study can be presented by the following equations:

$$P_i = a + b_1 L_k + b_2 E_j + \varepsilon \qquad \text{(Eq. 6.4)}$$

$$P_i = a + b_1 L_k + b_2 E_j + b_3 L_k E_j + \varepsilon \qquad \text{(Eq. 6.5)}$$

Where
P_i is the *i*th attribute for assessing performance improvement and $i = 1,2,3,4,5$
L_k is the *k*th attribute identifying the practice of intra-OL and $k = 1,2,3,4,5,6$
E_j is the *j*th attribute for evaluating the engagement in inter-OL and $j = 1,2,3,4,5,6,7,8$

A total of 240 MMR analyses, which are devised from the combination of the dependent variables (P_i), predictor variables (L_k and E_j), and moderator variables ($L_k E_j$) were conducted. The significance of the moderating effects was examined by both F-test and t-test.

Findings and discussions

Correlations between the engagement in inter-OL and practice of intra-OL

Table 6.5 shows the results of the Pearson correlation analysis. It is found that the contracting organisations' engagement in inter-OL in terms of 'addressing changes persistently (E_2)', 'revising the way of working (E_3)', 'seeking advice and suggestions from others (E_4)', 'revising long term strategy (E_6)' and 'the development of project monitoring system (E_7)' are significantly correlated with their practice of SL, DL and DeuL (L_1 to L_6) at $\rho < 0.05$ level. As such, the results support Hypothesis H1 as significant positive correlation is found between the practice of intra-OL and the engagement in inter-OL.

The findings are in agreement with that provided by Holmqvist (who pinpointed that organisations' practice of intra-OL is symbiotic to their engagement to learn at inter-organisational level). As such, the success of intra-OL is closely related to the contracting organisations' consciousness to learn from others (Holt *et al.* 2005).

Furthermore, among the attributes identifying engagement of inter-OL, the development of aq project monitoring system (E_7) is found to have a relatively stronger relationship with the practice of all OL types in terms of its correlation coefficient values (see Table 6.4). This echoes the findings as reported by Kasvi *et al.* (2003) who, with the case studies conducted, pinpointed that intra-OL can be fostered by the development of the project monitoring system. Based on other case studies conducted in Sweden, Huemer and Östergren (2000) also found that the performance records made available from a project monitoring system is a source of feedback from which lessons can be learnt. This further suggested that a project monitoring system in construction projects that provides regular and systematic feedback is an effective way to formalise the practice of inter-OL (Huemer and Östergren 2000).

Moderating effect of engaging in inter-OL

Out of 240 sets of MMR analyses conducted, the F-test and t-test results suggest that 29 of them showed significant moderating effects. To save space and preserve clarity, only the results that showed the significant moderating effects are presented in Table 6.5. As an illustration, the effect of 'doing things right in the first instance (L_1)' on performance improvement

Table 6.4 Correlations between the engagement in inter-OL and practice of intra-OL

| | Practice of intra-OL in terms of: | | | | | |
| | Single-loop learning (SL) | | Double-loop learning (DL) | | Deutero-learning (DeuL) | |
Engagement in inter-OL	Working (and considering corrective actions if required) under a clearly identified project goal (L_1)	Referring to the firm's past experience to interpret the performance feedback (L_2)	Identifying the root of the problem before taking improvement action (L_3)	Seeking and adopting new management and working approaches through evaluation of current practice (L_4)	Collecting and focusing on information which reflects the need to improve (L_5)	Systematising the communication channels among staff to ensure the awareness of the need to improve (L_6)
Seeking improvement methods from others (E_1)	0.19	0.19	0.38**	0.49**	0.20	0.54**
Addressing changes persistently (E_2)	0.29**	0.42**	0.35**	0.33**	0.50**	0.46**
Revising the way of working (E_3)	0.32**	0.40**	0.41**	0.47**	0.39**	0.50**
Seeking advice and suggestions from others (E_4)	0.23*	0.27**	0.29*	0.44**	0.28*	0.52**
Openness (E_5)	0.24*	0.21	0.32**	0.35**	0.09	0.33**
Revising long term strategy (E_6)	0.29**	0.35**	0.38**	0.49**	0.39**	0.56**
The development of project monitoring system (E_7)	0.45**	0.55**	0.49**	0.57**	0.54**	0.56**
The presence of incentive scheme (E_8)	0.13	0.27*	0.21*	0.44**	0.15	0.52**

Notes: *Significant level $p < 0.05$; **Significant level $p < 0.01$

Table 6.5 Results of the MMR analyses

Variables of the MMR analysis				(A)	(B)		(C)	
Dependent (Pi)	Predictor (L_k)	Moderator (E_j)	Predictor – Moderator product ($L_k E_j$)	ΔR^2	F-value for ΔR^2	Sig.	Std. coefficient of variable $E_j L_k$	Sig.
P_1	L_1	E_2	$L_1 X E_2$	0.07	6.80	*	0.34	**
		E_7	$L_1 X E_7$	0.05	5.54	*	0.34	*
	L_2	E_2	$L_2 X E_2$	0.05	6.09	*	0.27	*
		E_3	$L_2 X E_3$	0.04	6.85	*	0.28	*
		E_7	$L_2 X E_7$	0.01	3.72	+	0.16	*
	L_3	E_7	$L_3 X E_7$	0.04	3.26	+	0.22	*
	L_4	E_3	$L_4 X E_3$	0.05	8.77	**	0.28	*
		E_4	$L_4 X E_4$	0.06	5.19	*	0.14	*
		E_7	$L_4 X E_7$	0.04	5.75	*	0.27	*
P_3	L_1	E_2	$L_1 X E_2$	0.06	5.11	*	0.33	*
		E_3	$L_1 X E_3$	0.00	7.82	**	0.15	*
		E_7	$L_1 X E_7$	0.07	2.69	+	0.44	**
	L_2	E_2	$L_2 X E_2$	0.11	6.70	*	0.41	**
		E_3	$L_2 X E_3$	0.02	8.04	**	0.21	*
		E_7	$L_2 X E_7$	0.04	2.93	+	0.30	*
	L_4	E_4	$L_4 X E_4$	0.04	4.73	*	0.27	*
		E_5	$L_1 X E_5$	0.04	5.42	*	0.21	*
		E_6	$L_4 X E_6$	0.08	5.46	*	0.36	**
		E_7	$L_1 X E_7$	0.04	5.07	*	0.28	*
P_4	L_1	E_4	$L_1 X E_4$	0.07	4.29	*	1.39	**
	L_2	E_7	$L_2 X E_7$	0.11	2.94	+	0.46	**
	L_4	E_6	$L_4 X E_6$	0.09	3.21	*	0.37	**
P_5	L_1	E_7	$L_1 X E_7$	0.02	4.96	*	0.24	*
	I_2	E_2	$L_2 X E_2$	0.05	3.53	+	0.29	*
		E_4	$L_2 X E_4$	0.07	4.24	*	1.91	**
		E_7	$L_2 X E_7$	0.10	7.42	**	0.42	**
	L_4	E_3	$L_4 X E_3$	0.04	4.51	*	0.25	*
		E_6	$L_4 X E_6$	0.08	6.42	*	0.35	*
		E_7	$L_4 X E_7$	0.01	5.78	*	0.21	*

Remarks: $+ \rho <.1$, $*\rho <.05$, $\rho **<.01$ (one-tail)

Key:
P_1: Level of meeting the client's requirements
P_3: Being competent to change in order to meet the changing project requirements
P_4: Being competent to address forthcoming risks and consequences
P_5: Being competent to take prompt actions to tackle recurring problems
L_1: Working (and considering corrective actions if required) under a clearly identified project goal
L_2: Referring to the firm's past experience to interpret the performance feedback
L_3: Identifying the root of the problem before taking improvement action
L_4: Seeking and adopting new management and working approaches through evaluation of current practice
E_2: Addressing changes persistently
E_3: Revising the way of working
E_4: Seeking advice and suggestions from others
E_5: Openness
E_6: Revising long-term strategy
E_7: The development of a project monitoring system
E_8: The presence of incentive scheme

in terms of 'meeting the client's requirements (P_1)' is perceived to be contingent on the contracting organisations' engagement in 'addressing changes persistently (E_2)'. The ΔR^2 values obtained from the MMR analyses, as well as the results obtained from the respective F-tests and T-tests are reported in Columns A to C of Table 6.5.

The conceptual model as shown in Figure 6.1 proposed that the effect of practicing intra-OL on performance improvement is contingent on the contracting organisations' engagement in inter-OL. In other words, it proposed that if a contracting organisation engages in inter-OL (in terms of E_1 to E_8), the effect of practising of OL types (in terms of L_1 to L_6) on performance improvements (in terms of P_1 to P_5) becomes more significant. Nevertheless, the perceived moderating effects were only found in some of the results as shown in Table 6.5. The respective research findings are reported under three headings: inter-OL attributes, OL types and performance improvement with foci being placed on the more important observations. Table 6.6 presents such a summary.

Inter-OL attributes

From Table 6.5, it can be noted that some inter-OL attributes are more 'versatile' than the others. The term 'versatile' is used to describe an independent variable that is having significant effect on a wide range of dependent variables (Jaccard *et al.* 1990). In this respect, 'the development of a project monitoring system (E_7)' is the most versatile independent variable. This showed significant moderating effect on performance improvement in terms of the level of meeting the client's requirements (P_1),

Table 6.6 Summary of important observations

OL types that show good responses to a wide range of inter-OL attributes:
L_1: Working (and considering corrective actions if required) under a clearly identified project goal
L_2: Referring to the firm's past experience to interpret the performance feedback
L_3: Identifying the root of the problem before taking improvement action
L_4: Seeking and adopting new management and working approaches through evaluation of current practice
Inter-OL attribute which is effective in a wide range of situations:
E_7: The development of a project monitoring system
Performance improvements that are responsive to the intra-OL and inter-OL interactions:
P_1: Level of meeting the client's requirements
P_3: Being competent to change in order to meet the changing project requirements
P_5: Being competent to take prompt actions to tackle recurring problems

competence to change in order to meet with the changing project require-ments (P_3), competence to address forthcoming risk and consequences (P_4) and competence to take prompt actions to tackle recurring problems (P_5) when different OL types were practiced.

From a policing perspective, project monitoring systems (PMS thereafter) have been developed as a means to guard against inferior products (Wong and Cheung 2005). Moreover, PMS can also be identified as a system that records information reflecting project performance (Wong and Cheung 2005). It has been advocated that performance records can be a source of feedback from which lessons can be learnt (Huemer and Östergren 2000; Kurtyka 2003). For example, Robinson *et al.* (2004), based on case studies, pinpointed that the development of a project monitoring system can assist contracting organisations to create, acquire, capture and use knowledge, wherever it resides, to learn and improve. Furthermore, Ruuska and Vartiainen (2005) also supported that the development of a project moni-toring system among contracting organisations can facilitate inter-firm learning, which is vital for project performance improvement. Findings from this study suggest that the development of a PMS can be extended beyond policing and include providing feedback to contracting organisa-tions. Such feedback is vital for contracting organisations to practice intra-OL.

It is noted that these findings do not match those from previous studies, which identified that performance improvement is not a guaranteed result with the installation of a PMS (Love *et al.* 2004; Sense and Antoni 2003; Wong and Cheung 2005).This may be due to the fact that the conventional use of the PMS is for information recording and the attitude of contracting organisations is 'wait and see'. Thus PMS are incapable of energising contracting organisations to learn from mistakes (Love *et al.* 2004; Sense and Antoni 2003; Wong and Cheung 2005). This study timely highlights that the value of a PMS is on its capability to facilitate contracting organi-sations' learning from useful feedback.

OL types

Significant moderating effects were found when intra-OL is practiced in terms of 'working (and considering corrective actions if required) under a clearly identified project goal (L_1)', 'referring to the firm's past experience to interpret the performance feedback (L_2)', 'identifying the root of the prob-lem before taking improvement action (L_3)' and 'seeking and adopting new management and working approaches through evaluation of current prac-tice (L_4)'. As such, the above intra-OL practices characterise single-loop and double-loop learning of the contracting organisations. Furthermore, the results from the MMR analyses seem to indicate that for contracting organ-isations' the engagement in inter-OL may not significantly foster the effect of the practicing Deutero-learning (DeuL) on performance improvement.

Performance improvement

The MMR results suggest that interactions between the engagement of inter-OL and practice of OL types may not have a significant moderating effect on performance improvement in terms of assisting the contracting organisations to attain anticipated profit and address the forthcoming risk. Nevertheless, such interactions may foster performance improvement in terms of meeting the client's requirements (P_1), meeting with the changing project requirements (P_3), and taking prompt actions to tackle recurring problems (P_5). Indeed, the result is not surprising as rectifying the declining contracting organisations' performance and meeting the ever changing market demands have long been identified as the ultimate goal of organisational learning (Siriwardena and Kagioglou 2005).

In sum, the above findings support the proposition that practicing SL and DL have positive effect on performance improvement. In particular, such effect can be energised by the contracting organisations' engagement in learning from feedback provided by a PMS. Nevertheless, that not all PMSs as proposed in previous studies are developed purposively for energising contracting organisations' learning has been criticised (Kululanga *et al.* 2001; Love and Josephson 2004). The findings in this study are a timely reminder of the importance of designing a PMS with an OL dimension so as to make performance improvement sustainable.

Summary

This study investigates whether the effect of practicing intra-OL on performance improvement is contingent on the contracting organisations' engagement in inter-OL. It has been suggested from previous studies that the practice of intra-OL can be evaluated by the practice of three OL types: Single-loop learning (SL), double-loop learning (DL) and deutero-learning (DeuL). The results of this study suggest that contracting organisations' practice of SL and DL is symbiotic to their engagement to learn at inter-organisational level, in particular, from the feedback provided by a PMS. Furthermore, 'the development of the PMS' is identified as the most versatile inter-OL attribute in this study. It energises the effect of practising SL and DL on performance improvement in terms of meeting the project requirements, addressing the forthcoming risk and taking prompt actions to tackle recurring problems. The findings echo with previous studies, which advocated that a PMS that provides regular and systematic feedback can formalise the practice of inter-OL (Huemer and Östergren 2000). Nevertheless, construction researchers also noted the lack of attention given to facilitating contracting organisations' learning from mistakes by providing them with useful feedback (Sense and Antoni 2003). The findings in this study give evidence for the importance of developing a PMS purposively for vitalising contracting organisations' practice of learning.

Acknowledgement

The content of this chapter has been published in Volume 26(2) of the *International Journal of Project Management* and is used with the permission of Elsevier.

References

Aguinis, H. (1995) Statistical power problems with moderated multiple regression in management research. *Journal of Management,* 21(6), 1141–58.

Argyris, C. and Schön, D. (1978) *Organizational Learning: A Theory of Action Perspective.* Reading, MA: Addison-Wesley.

Bapuji, H. and Crossan, M. (2004) From questions to answers: reviewing organizational learning research. *Management Learning,* 35(4), 397–417.

Barlow, J. (2000) Innovation and learning in complex offshore construction projects. *Research Policy,* 29, 973–89.

Bresnen, M. and Marshall, N. (2000) Building partnerships: case studies of client–contractor collaboration in the UK construction industry. *Construction Management & Economics,* 18(7), 819–32.

Chan, P., Cooper, R. and Tzortzopoulos, P. (2005) Organizational learning: conceptual challenges from a project perspective. *Construction Management and Economics,* 23(7), 747–56.

Cheng, E. and Li, H. (2001) Development of a conceptual model of construction partnering. *Engineering Construction & Architectural Management,* 8(4), 292–303.

Choe, J. M. (2004) The relationships among management accounting information, organizational learning and production performance. *Journal of Strategic Information Systems,* 13, 61–85.

Cohen, J. and Cohen, P. C. (1983) *Applied Multiple Regression/Correlation Analysis for the Behavioral Sciences* (2nd cdn). Hillsdale, NJ: Lawrence Erlbaum.

Cohen, J., Cohen, P., West, S. G. and Aiken, L.S. (2003) *Applied Multiple Regression/Correlation Analysis for the Behavioral Sciences* (3rd edn). Mahwah, NJ: L. Erlbaum Associate.

Couto, J. P. and Teixeira, J. C. (2005) Using linear model for learning curve effect on highrise floor construction. *Construction Management & Economics,* 23(4), 355–64.

Cyert, R. M. and March, J. G. (1963) *A Behavioral Theory of the Firm.* Englewood Cliffs, NJ: Prentice-Hall.

El-Diraby, T. E. and Zhang, J. (2006) Constructability analysis of the bridge superstructure rotation construction method: the case of China. *Journal of Construction Engineering & Management,* 132(4), 353–62.

Enno, E. D., Koehn, P. E. and Datta, N. K. (2003) Quality, environmental, and health and safety management systems for construction engineering. *Journal of Construction Engineering and Management,* 129(9), 562–69.

Farghal, S. H. and Everett, J. G. (1997) Learning curves: accuracy in predicting future performance. *Journal of Construction Engineering and Management,* 123(1), 41–45.

Ford, D. N., Voyer, J. J. and Gould Wilkinson, J. M. (2000) Building learning organization in engineering cultures: case study. *Journal of Management in Engineering,* 16(4), 72–83.

Franco, L. A., Cushman, M. and Rosenhead, J. (2004) Project review and learning in construction industry: embedding a problem structuring method within a partnership context. *European Journal of Operational Research*, 152(6), 586–601.

Fu, W. K., Drew, D. S. and Lo, H. P. (2000) The effect of experience on contractors' competitiveness in recurrent bidding. *Construction Management and Economics*, 20(8), 655–66.

Groák, S. (1994) Is construction an industry? Notes towards a greater analytic emphasis on external linkages. *Construction Management and Economics*, 12, 287–93.

Hair, J. F., Anderson, R. E., Tatham, R. L. and Black, W. C. (1998) *Multivariate Data Analysis* (5th edn). Englewood Cliffs, NJ: Prentice Hall.

Holmqvist, M. (2003) A dynamic model of intra-and inter-organizational learning. *Organization Studies*, 24(1), 95–123.

Holt, G. D., Proverbs, D. and Love, P. E. D. (2000) Survey findings on UK construction procurement: is it achieving lowest cost, or value? *Asia Pacific Building and Construction Management Journal*, 5(2), 13–20.

Huemer, L. and Östergren, K. (2000) Strategic change and organizational learning in two Swedish construction firms. *Construction Management and Economics*, 18(6), 635–42.

Jaccard, J., Turrisi, R. and Choi, K. W. (1990) *Interaction Effects in Multiple Regression*. Thousand Oaks, CA: Sage

Jashapara, A. (2003) Cognition, culture and competition: an empirical test of the learning organization. *The Learning Organization*, 10(1), 31–50.

Kagioglou, M., Cooper, R., Aouad, G. and Sexton, M. (2000) Rethinking construction: the generic design and construction process protocol. *Engineering Construction & Architectural Management*, 7(2), 141–53.

Kasvi, J., Vartiainen, M. and Hailikari, M. (2003) Managing knowledge and knowledge competences in projects and project organizations. *International Journal of Project Management*, 21(8), 571–82.

Kululanga, G. K., McCaffer, R., Price, A. D. F. and Edum-Fotwe, F. (1999) Learning mechanisms employed by construction contractors. *Journal of Construction Engineering and Management*, 125(4), 215–33.

Kululanga, G. K., Edum-Fotwe, F. and McCaffer, R. (2001) Measuring construction contractors' organizational learning. *Building Research and Information*, 29(1), 21–29.

Kululanga, G. K., Price, A. D. F. and McCaffer, R. (2002) Empirical investigation of construction contractors' organizational learning. *Journal of Construction Engineering and Management*, 128(5), 385–91.

Kurtyka, J. (2003) Implementing business intelligence systems: an organizational learning approach. *DM Review Magazine*, November. Available at: www.dmereview.com/editorial/dmreview/print_action.cfm?articleId=7610

Law, K. M. Y. and Chuah, K. B. (2004) Project-based action learning as learning approach in learning organization: the theory and framework. *Team Performance Management*, 10(7/8), 178–86.

Li, H., Cheng, E., Love, P. and Irani Z. (2001) Co-operative benchmarking: a tool for partnering excellence in construction. *International Journal of Project Management*, 19(3), 171–79.

Love, P. and Josephson, P. (2004) Role of error-recovery process in projects. *Journal of Management in Engineering*, 20(1), 70–75.

Love, P. E. D., Huang, J. C., Edwards, D. J. and Irani, Z. (2004) Nurturing a learning organization in construction: a focus on strategic shift, organizational transformation, customer orientation and quality centred learning. *Construction Innovation*, 4, 113–26.

Murray, P. and Chapman, R. (2003) From continuous improvement to organizational learning: developmental theory. *The Learning Organization*, 10(5), 272–82.

Ofori, G. (2002) Singapore construction: moving towards a knowledge-based industry. *Building Research and Information*, 30(6), 401–12.

Pawlowsky, P. (2001) Management science and organizational learning, in M. Dierkes *et al.* (eds), *Handbook of Organisational Learning and Knowledge*. Oxford: Oxford University Press.

Pedler, M., Burgoyne, J. and Boydell, T. (1997) *The Learning Company: Strategies for Sustainable Development*. London: McGraw-Hill.

Redding, J. C. and Catalanello, R. F. (1994) *Strategic Readiness: The Making of Learning Organizations*. San Francisco: Jossey-Bass.

Robinson, H., Carrillo, P., Anumba, C. and Al-Ghassani, A. (2004) Developing a business case for knowledge management: the IMPaKT approach. *Construction Management and Economics*, 22(7), 733–43.

Ruuska, I. and Vartiainen, M. (2005) Characteristics of knowledge sharing communities in project organizations. *International Journal of Project Management*, 23(5), 374–79.

Sense, A. J. and Antoni, M. (2003) Exploring the politics of project learning. *International Journal of Project Management*, 21(7), 487–94.

Siriwardena, M. L. and Kagioglou, M. (2005) An integrative review of organisational learning research in construction, in *Proceedings of QUT Research Week 2005*, Queensland University of Technology, Australia. Available at: www.rics.org/NR/rdonlyres/5A225548-8EDD-487D-9F4E65EBAAB88B65/0/Integrative_review_organisational_learning20051129.pdf

Snell, S. A. and Dean, J. W. (1994) Strategic compensation for integrated manufacturing: the moderating effects of jobs and organizational inertia. *Academy of Management Journal*, 37(5), 1109–40.

Styhre, A., Josephson, P. and Knauseder, I. (2004) Learning capabilities in organizational networks: case studies of six construction projects. *Construction Management and Economics*, 22(9), 957–66.

Tjandra, I. K. and Tan, W. (2002) *Organisational Learning in Construction Firms: The Case of Construction Firms Operating in Jakarta, Indonesia*. Singapore: National University of Singapore.

Wong, P. S. P. and Cheung, S. O. (2005) From monitoring to learning: a conceptual framework, in *Proceedings of The 21st Annual Conference of the Association of Researchers in Construction Management*, SOAS, London, 1037–51.

Yiu, T. W. and Cheung, S. O. (2006) A study of construction mediator tactics – Part II: the contingent use of tactics. *Building and Environment*, 42(2), 762–69.

7 To learn or not to learn from project monitoring feedback

In search of explanations for the contractor's dichromatic responses

Peter Shek Pui Wong

Introduction

Research findings from the previous chapters suggested that the practice of the OL may lead to different degrees of performance improvement. Furthermore, the engagement in inter-OL is found to be vital for contracting organisations to practice single-loop learning (SL) and double-loop learning (DL).

Another soft power of a construction contracting organisation is the ability to respond effectively to changing client requirements. In particular, scholars emphasised that the construction contractors' practices should be reoriented to facilitate continuous improvement of their performance (Jashapara 2003; Kululanga *et al.* 1999; Love *et al.* 2000). In these studies, construction contractors have been typically described as a front-line workforce that converts construction project design into practical reality performance (Jashapara 2003; Kululanga *et al.* 1999; Love *et al.* 2000). Construction contractors are recognised as the hub of a construction supply chain as they not only link sub-contractors and suppliers, but also the client and the customers along the development process (Dainty *et al.* 2001). An unscrupulous contractor may deter project performance and lead to project failure (Wong 2004). Vice versa, an improved contractor's performance would increase client satisfaction as well as project value (Xiao and Proverbs 2003).

Such propositions have attracted several studies that focused on fostering contractors' learning in the construction projects (Al-JiBouri 2003; Love *et al.* 2004; Kululanga *et al.* 2002). Therefore, it makes good sense for researchers to explore ways to facilitate contractors' learning from past experience (Love *et al.* 2000; Wong and Cheung 2005). One way to achieve this, as suggested in a number of reported studies (Crawford and Bryne 2003; Franco *et al.* 2004; Orange *et al.* 2005), is to make use of feedback derived from project monitoring systems (PMS). In this study, PMS refers to a system that provides regular feedback to the construction contractors. In contractors' perspective, feedback derived from the PMS (described as project monitoring feedback hereafter) enables them to discover if they have achieved the pre-determined standards. In this regard, some researchers described project monitoring feedback as a vital learning resource (Franco *et al.* 2004; Kululanga *et al.* 1999). They demonstrated that when a contractor takes improvement action in light of feedback, an OL process occurs (Chan *et al.* 2005).

In this regard, various PMSs that provide different types of performance feedback for contracting organisations have been developed (see Table 7.1). These studies advocated that by providing performance feedback, contracting organisations' learning can be facilitated and subsequently performance

Table 7.1 Development of project monitoring systems (PMS) in construction

PMS	Descriptions
Workflow technology-based monitoring and control system (Shih and Tseng 1996)	A system applying network-based technology to track the flow of work and information, as well as the utilisation and commitment of resources along the project period.
Life cycle project management system (Chaaya and Jaafari 2001)	A system that evaluates the contracting organisations' achievement of the pre-agreed project goals.
Balanced score card performance evaluation system (Landin and Nilsson 2001)	A system that evaluates the contracting organisations' performance in financial, process, customer and learning perspectives. As such, financial and process evaluate the contracting organisations' past performance. Customer and learning evaluate contracting organisations' adaptability to the changes of customers' demands. A feedback loop is provided for effective communication about the demand changes among the contracting organisations, consultants and the clients.
Web-based construction project management system (Chan and Leung 2004)	A system developed under a web-based environment that enables interactive communication on the demarcation of the responsibility and the scope of work upon the changes of working requirements and orders. A bulletin board is developed in this system to enable on-line conferencing and e-mailing among all organisations in the construction supply chain.
Project performance monitoring system (Cheung *et al.* 2004)	The system evaluates the contracting organisations' performance by 8 major aspects, namely, people, cost, time, quality, safety and health, environment, client satisfaction and communication. The monitoring process is automated through the use of the World Wide Web and database technology. The automated monitoring process of PPMS affords the ease of set up and further adjustments of performance indicators to adapt to the change of the clients' demands.

improvement can be achieved (Ibbs *et al.* 2001; Franco *et al.* 2004). Based on a literature review, Orange *et al.* (2005) identified that the provision of performance feedback can stimulate contractors' learning. In a case study on the relationship between PMS and project performance, Franco *et al.* (2004) found contractors' learning from regular performance feedback to be imperative to performance improvement. In these studies, instigating improvement action has been coined as an outcome of OL.

Nevertheless, some researchers argued that OL is in fact not a guaranteed result with the installation of a PMS (Kululanga *et al.* 1999; Love *et al.* 2000; Choe 2004). Instead, contractors may only improve performance if they believed it is required (Kululanga *et al.* 1999). Interestingly, such arguments cohere with the findings reported in the last chapter indicating the close relationships among turbulence, unlearning and organisational learning.

In previous research studies, PMSs were developed to foster contractor's OL (Crawford and Bryne, 2003; Dikmen *et al.* 2005; Franco *et al.* 2004). Contractor was often postulated as a rational entity who responds optimally to the project monitoring feedback (Everett and Farghal 1997; Love and Josephson 2004). Researchers seemingly assumed an exclusive relationship between the provision of project monitoring feedback and the contractor's practice of OL (Everett and Farghal 1997; Franco *et al.* 2004). The extent of a contractor's learning may also be dependent on how frequently the project monitoring feedback is generated and received (Everett and Farghal 1997). However, should the contractors learn and consequently take improvement actions simply because they were provided project monitoring feedback? Negative views were found from some recent studies (Martin and Root 2010; Wong *et al.* 2010). Based on the results obtained from a survey conducted in South Africa, Martin and Root (2010) found that the contractors did not capitalise on the project monitoring feedback to take improvement actions as the developers desired. A case study conducted by Lau and Rowlinson (2011) revealed that equity and fairness are not entrenched in the client-contractor relationship and thus inhibit OL. In this regard, Wong *et al.* (2010) empirically tested the effect of project monitoring feedback on the contractors' performance. They discovered that many contractors cease to improve when they reach a performance level that meets the client's minimum requirement. As such, the above findings suggest that perhaps contractors' learning is a subtle process that embodies the 'discontinuous transitions of behaviour' (Raijmakers *et al.* 1996: 105). The situation is similar to ice melting, which involves the transition from solid to liquid phase of water. To take or not to take improvement action is possibly a discontinuous decision that may be associated with the contractor's expectation of the performance (Wong *et al.* 2010). In other words, a sudden change in learning behaviour may occur when the contractors' performance surpass (or in vice versa fall below) expectation (Wong *et al.* 2010). Regrettably, empirical studies focused on such sudden transition of learning behavior remain scant (Wong *et al.* 2008). This may be because such behaviour entails the concept of discontinuity, which is not possible to be mathematically presented by linear models in the same way as multiple regression or structural equation models (Cheung *et al.* 2008; Dou and Ghose 2006).

In this aspect, the use of non-linear models to conceptualise organisational behaviour change is well documented in other research fields (Dou and Ghose 2006; Garud and Van de Ven 1992). Garud and Van de Ven (1992) emphasised that the use of linear and non-linear models to display an organisation's decision outcome is not mutually exclusive. However, a linear model may not offer a good explanation on a decision made by an organisation during an ambiguous and uncertain period of time. They utilised catastrophe theory to capture the non-linear dynamics of the internal corporate venturing decision. A similar approach was adopted by Dou and Ghose (2006) who captured the inherent non-linearity and complexity of the online retail competition. They found that if the visitors' hit-rates between the rival online retail pages progressively change in the opposite direction, the customer-base can suddenly shift from one company to another. Catastrophe theory was used to provide a theoretical explanation of the sudden behavioural change. Following this stream of studies, it is noted that catastrophe theory may be applicable for explaining the change of the contractor's learning response to the project monitoring feedback.

This chapter reports a study that aims to examine the inherent dynamics among: (1) the contractors' practice of OL; (2) their attention to their project monitoring feedback; and (3) their anticipated performance in a project by using the catastrophe theory. It posits that the contractor's learning from project monitoring feedback can be triggered by their anticipation of the project performance and can be modelled as a sudden attitudinal change. The following hypothesis is thus proposed:

H1: The contractors' practice of OL (OL) in response to their attention to project monitoring feedback (AF) is contingent on their anticipation of project performance (PP) in the construction projects.

This study offers a different approach to model the relationship between OL and performance improvement actions. This may assist in unveiling the tactics utilised by the contractors in light of regular project monitoring feedback. Learning behaviours of the contractors are studied because, amongst those construction organisations, contractors are responsible for converting building designs into physical facilities. Their learning capabilities and decisions thus would directly affect project success and the competitiveness of the entire construction supply chain (Sense and Antoni 2003; Xiao and Proverbs 2003).

The rest of this chapter is organised as follows: The second section provides an introduction to catastrophe theory. Based on catastrophe theory, a conceptual model is developed depicting the hypothesised relationships among: (1) contractors practice of OL; (2) their attention to project monitoring feedback; and (3) their anticipated performance in a project. The third section describes the methodologies employed for examining the hypothesised relationships. The fourth section reports and discusses the implications of the findings.

Catastrophe theory

Catastrophe theory is a mathematical model of non-linear relationships developed by Thom (1975) and subsequently popularised by Zeeman (1977). It expresses an abrupt transition of equilibrium behaviour (i.e. the dependent variable) which is caused by a smooth gradual change of the control factors (i.e. the independent variables). Thom described catastrophe theory as a hypothesis of structural stability. He hypothesised that inherent stability (i.e. a state of equilibrium) of a system can be maintained by a set of parameters which values are changing over time. However, such inherent stability may not be sustained when a particular value or values of the parameters reach a threshold. Consequently, the state of equilibrium of the system would undergo a sudden change. In other words, the discontinuity in behaviour can be modelled as a function of a progressive change of values of a set of parameters (Grasman *et al.* 2009; Thom 1975; van der Maas *et al.* 1992, 2003). Stamovlais and Tsaparlis (2012) applied catastrophe theory in the field of education science. They described catastrophe theory as a concept that can provide quantitative-based explanations for the sudden behavioural change in student learning. Dou and Ghose (2006) described catastrophe theory as a tool for modelling operational systems the inner workings of which may not be effectively depicted by using linear models. Under this logic, catastrophe theory may be considered appropriate for modelling the inherent complexity of the contractor's behaviour. In the construction research field, Yiu and Cheung (2006) conceptualised conflict behaviour of the construction professionals by using a cusp model. Cusp is one of the simplest models in the family of catastrophes (van der Maas *et al.* 2003). A typical cusp model consists of one dependent variable and two independent variables. The two independent variables (namely normal factor and splitting factor) carry different qualitative meanings. The normal factor 'is related to the dependent variable in a consistent pattern' (Yiu and Cheung 2006: 439). The splitting factor is a 'moderator which specifies conditions under which the normal factor will affect the dependent variable in a continuous fashion, and other circumstances under which the normal factor will produce discontinuous changes in the dependent variable ... it is the splitting factor that determines the breaking point or threshold of change in the dependent variable' (ibid.). Yiu and Cheung hypothesised conflict as the dependent variable, tension as the normal factor and behavioural flexibility as the splitting factor. A questionnaire survey was conducted in Hong Kong, and the respondents were asked to provide their perceptive views on the conflict, tension and behavioural flexibility by using a seven point Likert scales. Yiu and Cheung found that neither the linear nor the logical model can provide a statistically valid explanation to the inter-relationships among conflict, tension and behavioural flexibility. However, catastrophe model is found statistically fit for describing such relationship. The results indicate a sudden jump in conflict level will occur when tension between the construction professionals reaches the threshold (Yiu and Cheung 2006). Another study conducted by Cheung *et al.* (2008) also applied cusp model to examine the contracting behaviour of the contracting parties in the construction projects. It was found that while contracting party's dissatisfaction on the counterpart intensifies, it remains cooperative up to one point beyond which it will suddenly fight back by taking aggressive responses in the project. The latest study conducted by Chow *et al.* (2012) successfully adopted the cusp model to demonstrate how parties can avoid sudden withdrawal from project dispute negotiations. The above studies suggested that linear and logistic regression models may not give a good explanation for an individual's or organisation's sudden change of behaviour. In this situation, the catastrophe theory may provide a promising framework for modelling such behaviour. Indeed, the cusp model has widely been applied with great diversity of management studies such as the organisational decision-making process (van der Maas *et al.* 2003), technology management (Herbig 1991) and customer behaviour (Oliva *et al.* 1981). In this study, the contractors' practice of OL conceptualised as a dependent variable in the cusp model in which their attention to project monitoring feedback serves as a normal factor and anticipated performance in the projects serve as a splitting factor. The normal and splitting factors are conceptualised to affect the dynamic of the practice of OL. As such, the conceptual cusp model is underpinned by three streams of studies: practice of OL, attention to feedback and anticipation of project performance.

The conceptual model

Practice of OL

The current study benefited from this wealth of research reported in the previous chapters and defines the contractors' OL as a process through which an organisation imbibes and applies the acquired knowledge for performance improvements (Kululanga *et al.* 1999). Regarding the contractors' practice of OL, this study considers project monitoring feedback as a learning source that is similar to the definition of deutero-learning (DeuL). Findings reported in Chapter 5 indicate DeuL may not work as a learning type but as a platform that facilitates the achievement of SL and DL. In this regard, in this study contractors' OL are evaluated by their practices of SL and DL. The attributes that are used for identifying the practice of SL and DL are in line with those reported in Chapter 5 and 6.

Attention to project monitoring feedback (AF)

The second stream of literature is related to the attention to project monitoring feedback. Researchers suggested that project monitoring feedback is provided to the contractors to: (1) spot out non-compliance; (2) avoid continuous faulty actions; and (3) facilitate resources reallocation (Wong *et al.* 2010; Franco *et al.* 2004). The purpose of project monitoring feedback is to make the contractors aware that the delivery of the pre-determined project standards is their common goal (Brown and Adams, 2000). As such, project monitoring feedback is regarded as a source that draws the contractor's attention to minimise the deviations between actual and predetermined standards (Love and Josephson 2004). A construction project has conventionally been viewed as successful if it came within the budget, original schedule and quality of work. In this connection, it is not surprising that time, cost and quality are the most common areas on which the contractors regularly receive project monitoring feedback from the developers and consultants.

Others argued that time, cost and quality merely represent project performance in terms of efficacy, without due regard to the importance of effectiveness (Toor and Ogunlanga 2010; Wong *et al.* 2008). Project efficacy refers to success of attaining the pre-determined goals (Wong *et al.* 2008). In a contrary, project effectiveness concerns the capability of accomplishing the goals. In recent years, various scholars have made suggestions about how project effectiveness can be evaluated (Kagioglou *et al.* 2001). Additional metrics such as how well the contractor communicated with the project manager or how well resources are allocated by the contractor to complete the project task were suggested to expand the scope of project monitoring feedback (Crawford and Bryne 2003). Key performance indicators (KPIs) were established to offer a wide range of 'objective criteria to measure the success of a project' (Toor and Ogunlanga 2010: 229). In other words, KPIs were developed for providing feedback to contractors in different angles so as to improve the robustness of project monitoring (Toor and Ogunlanga 2010; Crawford and Bryne, 2003; Kagioglou *et al.* 2001). Other influential work related to project monitoring feedback is the Balanced Scorecard (BSC) framework developed by Kaplan and Norton (1992). A number of studies have used the BSC framework to develop PMSs specifically for construction projects (Kagioglou *et al.* 2001, Mohamed 2003). For example, Kagioglou *et al.* (2001) developed PMSs that provide performance feedback in four perspectives: financial, internal business processes, customer and innovation. Mohamed (2003) applied the BSC approach to develop construction safety performance indicators. He suggested that feedback on construction safety performance can be provided in four strategic dimensions: (1) management, (2) operation, (3) customer, and (4) learning. The above studies suggested that construction researchers have endeavoured to widen the scope of project monitoring. While the use of additional metrics have been promoted, researchers never ruled out time, cost and quality in their proposed list of project monitoring feedback. With due caveats about the fact that there can be various types of project monitoring feedback, this study concentrates on the contractor's attention to the three most common types of feedback: time, cost and quality.

For the purpose of this study, it is necessary to devise operational statements to characterise the contractor's attention to feedback. The related operational statements are listed in Table 7.2.

Anticipation of project performance (PP)

Sceptics have long challenged the relationship between the provision of project monitoring feedback and the contractor's learning. Based on the results obtained from a survey conducted in United Kingdom, Dikmen *et al.* (2005) found that the contractors' performance was improved significantly, notwithstanding that they had been provided with sufficient feedback from the developers. A case study conducted by Love *et al.* (2000) also pinpointed that contractors rarely take full advantage of performance feedback in addressing improvement actions. While pinpointing that performance feedback helps the contractors to learn from the past, Love and Josephson (2004) argued that the contractors may not learn from feedback amicably and continuously. Unless vulnerable to negative feedback, contractors rarely take improvement action (Greve 2002; Wong *et al.* 2010). Similarly, scholars from the field of organisational management have long pinpointed that organisations typically operate under constraints such

Table 7.2 Typical project monitoring feedback obtainable from a PMS

Project monitoring feedback types	Operational Statement: My firm is paying attention to the following project monitoring feedback received from the client/his consultants:	*References*					
		A	*B*	*C*	*D*	*E*	*F*
Time (AF1)	Actual project progress as compared to the original programme	*	*	*	*	*	*
Cost (AF2)	Actual project expenditure as compared to the original cost plan	*	*	*	*		*
Quality (AF3)	Actual project quality as compared to the project quality stipulated in the specification	*	*	*			*

References:
(A) Brown and Adams 2000
(B) Kagioglou *et al.* 2001
(C) Crawford and Bryne 2003
(D) Mohamed 2003
(E) Love and Josephson 2004
(F) Wong *et al.* 2012

as ill-defined goals, limited time and budgets, intangible rewards driven from taking improvement actions, and so forth (Harrison and Palletier 1995; Simon 1955). As long as not breaching the contract terms, organisations are not amenable to work for the best, despite such behaviour being desired by their counterparts (Harrison and Palletier 1995). Such behaviour has been described by a term 'satisficing' which was first articulated by Simon (1955) who coined this as a portmanteau of two words, satisfy and suffice (Harrison and Pelletier 1995; Simon 1955). Simon argued that in the real world, organisations would likely accept the constraints and as a result outcomes are suboptimal. Instead of searching for an optimal solution, they typically cease the search process when they find an alternative solution that can satisfy their expectation. Extending the work of Simon, Harrison and Pelletier (1995) described in essence, satisficing behaviour occurs when the managerial decision maker sets up a feasible level of aspiration and then searches for alternatives until finding one that achieves the level. As soon as a satisficing alternative is found, the search stops and the decision maker proceeds toward implementation of the satisficing course of action.

The above literature review indicates that the contractor's learning response to feedback may suddenly cease when their performance meets expectation. In this connection, it becomes logical to question whether a threshold of performance level resides in the contractors' mind beyond which the learning behaviour would change instantaneously.

In this study, 'anticipated performance' refers to the contractor's anticipation of his performance as compared to the three pre-determined project goals: time, cost and quality. An anticipated performance does not necessarily reflect the actual performance (Harrison and Pelletier 1995). However, previous studies suggested that anticipated performance may determine how the contractor will perform in the project (Cheung *et al.* 2008; Harrison and Pelletier 1995). For the purpose of this study, contractors were asked to evaluate their performance in attaining the client's pre-determined targets in time, cost and quality as compared to their original anticipation.

The conceptual model

Based on the discussion in the foregoing three sections, a relationship framework among OL, AF and PP arranged in a cusp model format is proposed (Figure 7.1). Generally, it is hypothesised that the contractors' practice of OL (OL) in response to their attention to feedback (AF) is contingent on their anticipation of project performance (PP) in the construction projects. To this end, the contractors' practice of OL is hypothesised as a dependent variable in the cusp model in which their attention to project monitoring feedback is hypothesised as a normal factor and anticipated performance a splitting factor. The model can be further described by the following three catastrophe phenomenon: (1) bimodality, (2) sudden jump, and (3) hysteresis (Gilmore 1981).

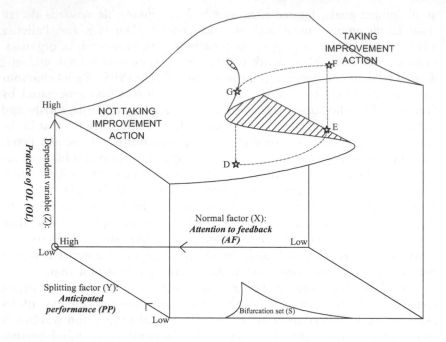

Figure 7.1 The conceptual model

Bimodality

Bimodal means that the 'practice of OL' level may be either high or low when the contractor's level of attention to feedback is moderate. If the contractor anticipates that his performance will surpass the client's expectation, he tends to resist practising OL until the 'attention to feedback' levels take an opposite extreme value. The thresholds of the low or high level of 'attention to feedback', which demarcate the switching points from 'not taking improvement action' (a state whereby anticipated performance meets or surpasses client's expectation and low level of attention to feedback) to 'taking improvement action' (a state whereby anticipated performance is lower than client's expectation and high level of attention to feedback) and vice versa, emerge (Yiu and Cheung 2006).

Sudden jump

Contractors have been criticised for their inadequate learning action in response to feedback. Such a statement can be explained by the organisational learning behaviour as described by Harrison and Pelletier (1995).

Harrison and Pelletier explained that contractors typically operate with: (1) ill-defined goals; (2) limited time and budgets; and (3) lack of motives to improve. As long as organisations are not breaching the contractual terms, they are not amenable 'to work for the best'. In other words, they normally prefer working towards a tenable performance standard that is sometimes lower than their attainable best even if they have been driven by their clients. The sudden jump (Path EF in Figure 7.2) indicates that the practice of OL level suddenly rises when the contractor anticipates that they may perform poorer than the client's expectation while their level of attention to feedback remains moderate.

Hysteresis

Anticipated performance not only can trigger but also create resistance to the practice of OL. When the contractor anticipates that his performance will meet or surpass the client's expectation, a counter reaction to the practice of OL occurs. This phenomenon (as demonstrated in path EFGD in Figure 7.2) is called hysteresis.

Research methodologies

Measurement of construct and questionnaire

To test the conceptual model, a questionnaire survey was conducted for data collection. The questionnaire contains four parts. Part 1 deals with demographic information about the respondents. Respondents were asked to specify a project in which they have been participating for at least one

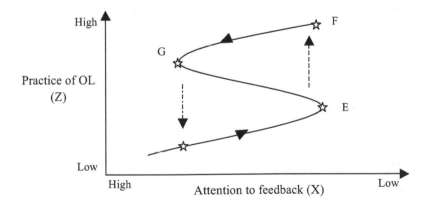

Figure 7.2 Hysteresis path 'DEFGD' as shown in Figure 7.1

year, and the questionnaires of those not having taken part in a specified project for more than one year were discarded. The respondents were requested to answer the questions in Parts 2 to 4 of the questionnaire by referring to their experience gained in the specified project. In Part 2, respondents were asked to evaluate the extent to which the client's expectation about the project time, cost and quality can be met. Part 3 seeks to solicit the degree of respondents' agreement on their company's attention to the project monitoring feedback (see Table 7.1). In Part 4, questions engage the extent of SL and DL being practised by the respondent's firm in response to the three types of project monitoring feedback. A pilot study that involved 12 industry experts was conducted before undertaking the full-scale survey to ensure the relevance of questions to the construction project scenario. A sample of the questionnaire is presented in Figure 7.3.

Response rate and sample profiles

Of the total responses received, 113 were used representing a 38 per cent response rate. This outcome is comparable with questionnaire surveys of this kind (Cheung *et al.* 2008; Yiu *et al.* 2007). The study has attracted a reasonable response rate in comparison to other questionnaire surveys in the construction field, normally ranging from 25 per cent to 30 per cent (Wong *et al.* 2012). Likewise, the response rate of the current research is similar to that of the study related to catastrophic transitions of construction contracting behaviour by Cheung *et al.* (2008) (i.e. 91 responses out of 250 questionnaires being sent out or 36 per cent of the response rate). The sample size and the return rate of this survey study are considered acceptable.

It should be noted that more than 75 per cent of the respondents have more than 10 years of project management experience (Table 7.3). The creditability of the respondents is indicative of their service to the industry thus their responses are considered to be a reflection of the industry's views.

Table 7.3 Respondent's working experience

Working experience	No. of respondents	%
Below 5 years	8	7
5–10 years	19	17
11–15 years	33	29
16–20 years	30	27
Over 20 years	23	20
Total	113	100

Part 1- Demographic Information					
Q1.1	Your working experience in the construction field:	a:	<5 years	b:	5-10 years
		c:	11-15 years	d:	16-20 years
		e:	>20 years		
Q1.2	Your position in the company:	a:	Project Manager	b:	Engineer
		c:	Quantity surveyor	d:	Project coordinator
		e:	Clerk of works	f:	Others (pls. specify)
Q1.3	No. of employees in your company	a:	<20	b:	21-50
		c:	51-100	d:	>100

With reference to one construction **project** that you have been involving for at least 1 year and provide the following particulars:

Q1.4	Project nature	a: Residential	b: Office/Amenities		c: Office/Amenities
Q1.5	Project Sum ($,000s) (optional)	a:	<5,000	b:	5,000-20,000
		c:	>20,000		
Q1.6	Project Name :				

Part 2- Project Performance
The recent performance of your company (in the following aspects) in the *project* stated in Q1.6 is:
(1: Worse than the client's expectation ↔ 4: Meeting the client's expectation ↔ 7: Better than the client's expectation)

Q2.1	Time (for example: Have recent project tasks been completed within client's schedule?) **(PP1)**	1 2 3 4 5 6 7
Q2.2	Cost (for example: Have recent project tasks been completed within client's budget?) **(PP2)**	1 2 3 4 5 6 7
Q2.3	Quality (for example: Have the recent completed project tasks complied with the Specification requirement?) **(PP3)**	1 2 3 4 5 6 7

Part 3- Attention to feedback
Based on your assessment in Part 2, do you agree that your firm has paid attention to the following project monitoring feedback received from the client/his consultant? : **(1 = Disagree strongly ↔ 7 = Agree strongly)**

Q3.1	Actual project progress as compared to the original programme **(AF1)**	1 2 3 4 5 6 7
Q3.2	Actual project expenditure as compared to the original cost plan **(AF2)**	1 2 3 4 5 6 7
Q3.3	Actual project quality as compared to the project quality stipulated in the Specification **(AF3)**	1 2 3 4 5 6 7

Part 4- Practice of organisational learning
Upon receiving the following project monitoring feedback received from the client/his consultant, my firm: **(1 = Disagree strongly ↔ 7 = Agree strongly)**

		Feedback in time (as stated in Q3.1)	Feedback in cost (as stated in Q3.2)	Feedback in quality (as stated in Q3.3)
Q4.1	Works (and considers corrective actions if required) under a set clearly identified project goal **(SL1)**	1 2 3 4 5 6 7	1 2 3 4 5 6 7	1 2 3 4 5 6 7
Q4.2	Refers to its experience to interpret the performance feedback **(SL2)**	1 2 3 4 5 6 7	1 2 3 4 5 6 7	1 2 3 4 5 6 7
Q4.3	Identifies the root of the problem before taking improvement action **(DL1)**	1 2 3 4 5 6 7	1 2 3 4 5 6 7	1 2 3 4 5 6 7
Q4.4	Seeks and adopts new management and working approach through evaluation of current practice **(DL2)**	1 2 3 4 5 6 7	1 2 3 4 5 6 7	1 2 3 4 5 6 7

Figure 7.3 The sample questionnaire

Data analysis method

To test the hypothesised relationships among OL, AF and PP, a catastrophe model fitting technique was adopted. Catastrophe model fitting technique is a statistical technique that is based on the stochastic differential equations to estimate model parameters from the data (Cobb and Zacks 1985). The conceptual cusp model in its mathematical form can be expressed by the following equations:

$$V_{\alpha\beta}(z) = \tfrac{1}{4}z^4 + \tfrac{1}{2}\beta z^2 + \alpha z \qquad\qquad\text{(Eq. 7.1)}$$

where β and α are the linear functions that represent the bifurcation and the asymmetry of the set of independent variables X (attention to feedback) and Y (anticipation of performance), z is the probability density function for setting the first variate of Equation 7.1. Critical change of behaviour occurs when $V_{\alpha\beta}(z) = 0.$, that is

$$z^3 + \beta z + \alpha = 0 \qquad\qquad\text{(Eq. 7.2)}$$

where

$$z = (S - \lambda)/\sigma \qquad\qquad\text{(Eq. 7.3)}$$

In Equation 7.3, z is the probability density function, S represents the dependent variable (practice of OL), λ is an estimate of the location of S and σ is the standard deviation of S. In the cusp model, the asymmetry (α) and the bifurcation (β) of the set of independent variables (i.e. X: attention to feedback and Y: anticipation of performance) are denoted by:

$$\alpha = a_o + a_X X + a_Y Y \qquad\qquad\text{(Eq. 7.4)}$$

$$\beta = b_o + b_X X + b_Y Y \qquad\qquad\text{(Eq. 7.5)}$$

where
X represents Attention to feedback (AF)
Y represents Anticipation of Performance (PP)
a_o and b_o represent the intercepts of the asymmetry (α) and the bifurcation (β) linear functions
a_X, a_Y, b_X and b_Y represent the regression coefficients

Based on the above equations, Cobb (1980) suggested that parameters λ, σ, a_o, a_X, a_Y, b_o, b_X and b_Y can be estimated by using a maximum likelihood method (Cobb 1980; Cobb and Zacks 1985; Van der Maas *et al.* 2003; Yiu and Cheung 2006). As such, the loadings of the control parameters can be computed by using the above equations (Cobb 1980; Cobb and Zacks 1985; Yiu and Cheung 2006).

The maximum likelihood method proposed by Cobb (1980) has been described as the most statistically satisfactory method for fitting the fitness of the conceptual cusp model (Cobb and Zacks 1985; Van der Maas *et al.* 2003; Yiu and Cheung 2006). Notwithstanding, the computer program designed by Cobb (1980) based on his proposed model fitting technique was not often used. Scholars suggested that this may be due to the program often breaking down for no apparent reason (Van der Maas *et al.* 2003; Yiu and Cheung 2006). Extended from the work of Cobb (1980), Hartelman (1997) developed a computer program called Cuspfit to facilitate the catastrophe model fitting. Hartelman (1997) and Wagenmakers *et al.* (2004) suggested that Cuspfit is a more robust and flexible version than Cobb's original program. It employs a more reliable optimisation routine, which allows users to test the fitness of using a cusp model to describe the relationship among the variables in the conceptual model (Cheung *et al.* 2008; Cobb 1980; Wagenmakers *et al.* 2004; Yiu and Cheung 2006).

Hartelman (1997) introduced two fit measures in Cuspfit: Akaike Information Criterion (AIC) and Bayes Information Criterion (BIC). 'AIC is the goodness-of-fit index that takes account of the number of parameters. Mathematically, it is defined as minus twice the loglikelihood plus twice the number of parameters, i.e. AIC $=-2 \log L+2k$; the model with the smallest AIC will be the best fit' (Yiu and Cheung 2006: 442). 'BIC is a goodness-of-fit indicator that takes into account the number of data points and implements Occam's razor by quantifying the trade-off and parsimony' (ibid.). BIC can be computed by an equation BIC$=-2 \log L+k \log n$ where L=maximum likelihood; k=number of free parameters; and n=number of observations (Cobb 1980). Models with lower BIC values are preferred for model fitness purposes (Yiu and Cheung 2006; Cheung *et al.* 2008). As such, the lower the AIC and BIC values, the better the fitness of using the cusp model to describe the relationship between the behaviour variable and the control variables (Cheung *et al.* 2008; Cobb 1980; Wagenmakers *et al.* 2004; Yiu and Cheung 2006).

Cuspfit not only enables the ability to test the fitness of the proposed cusp model, but also the logistic and linear models to describe the relationship among the variables of the conceptual models (Cheung *et al.* 2008; Cobb 1980; Wagenmakers *et al.* 2004; Yiu and Cheung 2006). With this function, Cuspfit can be used to test the presence of bifurcations by comparing the fit of the cusp model to the fit of both logistic and linear models in terms of their AIC and BIC values (Cheung *et al.* 2008; Hartelman 1997; Yiu and Cheung 2006). Cuspfit has been successfully applied in a wide range of studies in management, social and behavioural science research to test the existence of a catastrophe relationship (Cobb 1980; Van der Maas *et al.* 2003; Yiu and Cheung 2006). The method has also been applied in construction research. For instance, Yiu *et al.* (2007) applied Cuspfit to depict the existence of a dichotomise pair of cooperation and aggression forces in the construction contracting environment.

This study employed Cuspfit to examine whether the conceptual model is statistically significant to be explained by the cusp model. For a fair comparison of the results, the fitness of the relationships among OL, AF and PP to the linear and log-linear models was also examined. Twelve sets of variables (as illustrated in Table 7.4) were tested by using Cuspfit.

Table 7.4 Combinations of variables for Cuspfit

Variables Set	Practice of OL (Z)	Attention to feedback (X)	Anticipation of project performance (Y)
1	SL1	AF1	PP1
2	SL2	Ditto	Ditto
3	DL1	Ditto	Ditto
4	DL2	Ditto	Ditto
5	SL1	AF2	PP2
6	SL2	Ditto	Ditto
7	DL1	Ditto	Ditto
8	DL2	Ditto	Ditto
9	SL1	AF3	PP3
10	SL2	Ditto	Ditto
11	DL1	Ditto	Ditto
12	DL2	Ditto	Ditto

Results and discussions

By using the Cuspfit program, the AIC and BIC values of the cusp, logistic and linear models for the twelve variables sets were computed. The results are presented in Table 7.5.

The results as shown in Table 7.5 indicate that for Variable Sets 1, 2, 5, 6, 11 and 12 the linear model fits the data sets with the lowest loglikelihood, AIC and BIC values. Either SL1 or SL2 was used as the dependent variables in these Variable Sets. Interestingly, 'SL1: working (and considering corrective actions if required) under a clearly identified project goal' and 'SL2: referring the firm's experience to interpret the performance feedback' characterise the practice of single-loop learning (SL). The above findings in a broad sense suggest the positive linear relationships among the practice of SL, the contractor's attention to feedback (AF) and their anticipation of performance (PP). The results echo the findings of Murray and Chapman (2003) that learning from project monitoring feedback is vital for contractors to devise performance improvement actions. Some sceptics argued that contractors may not be willing to learn from project monitoring feedback unless lack of improvement action would result in further sanctions (Love and Josephson 2004; Wong *et al.* 2010). The results of this study indicate that contractors may practice SL although their performance may have already surpassed the client's expectation. One possible reason for this result is that SL enables alteration of actions without scrutinising the underlying assumptions, leading to the difference between the expected and the actual outcomes. In this sense, SL seems to be an easier option for a quick fix, whereas DL practices call for the re-examination of some, if not all, of the fundamentals.

Table 7.5 Analysis results for fitting data to cusp, logistic and linear models

	Log likelihood	AIC	BIC
Variables Set 1			
Cusp	−0.13E+03	0.25E+03	0.26E+03
Logistic	−0.15E+03	0.30E+03	0.31E+03
Linear	**−0.11E+03**	**0.24E+03**	**0.26E+03**
Variables Set 2			
Cusp	−0.12E+03	0.24E+03	0.26E+03
Logistic	−0.15E+03	0.31E+03	0.32E+03
Linear	**−0.11E+03**	**0.23E+03**	**0.25E+03**
Variables Set 3			
Cusp	**−0.12E+03**	**0.24E+03**	**0.26E+03**
Logistic	−0.15E+03	0.31E+03	0.32E+03
Linear	−0.14E+03	0.26E+03	0.28E+03
Variables Set 4			
Cusp	**−0.11E+03**	**0.24E+03**	**0.25E+03**
Logistic	−0.15E+03	0.30E+03	0.31E+03
Linear	−0.15E+03	0.25E+03	0.26E+03
Variables Set 5			
Cusp	−0.13E+03	0.24E+03	0.28E+03
Logistic	−0.16E+03	0.33E+03	0.32E+03
Linear	**−0.10E+03**	**0.19E+03**	**0.22E+03**
Variables Set 6			
Cusp	−0.14E+03	0.26E+03	0.26E+03
Logistic	−0.19E+03	0.28E+03	0.30E+03
Linear	**−0.10E+03**	**0.20E+03**	**0.22E+03**
Variables Set 7			
Cusp	**−0.10E+03**	**0.18E+03**	**0.20E+03**
Logistic	−0.18E+03	0.30E+03	0.29E+03
Linear	−0.15E+03	0.23E+03	0.22E+03
Variables Set 8			
Cusp	**−0.12E+03**	**0.20E+03**	**0.19E+03**
Logistic	−0.16E+03	0.31E+03	0.29E+03
Linear	−0.15E+03	0.23E+03	0.21E+03
Variables Set 9			
Cusp	−0.15E+03	0.22E+03	0.23E+03
Logistic	−0.18E+03	0.29E+03	0.30E+03
Linear	**−0.12E+03**	**0.18E+03**	**0.22E+03**
Variables Set 10			
Cusp	−0.13E+03	0.25E+03	0.23E+03
Logistic	−0.16E+03	0.30E+03	0.28E+03
Linear	**−0.10E+03**	**0.22E+03**	**0.18E+03**
Variables Set 11			
Cusp	**−0.10E+03**	**0.18E+03**	**0.15E+03**
Logistic	−0.17E+03	0.30E+03	0.31E+03
Linear	−0.15E+03	0.25E+03	0.26E+03
Variables Set 12			
Cusp	**−0.12E+03**	**0.22E+03**	**0.20E+03**
Logistic	−0.20E+03	0.32E+03	0.29E+03
Linear	−0.16E+03	0.26E+03	0.22E+03

The result is consistent with previous findings about the way to meet the client's predetermined project targets; that contractors prefer practicing SL (Kululanga *et al.* 1999, Wong *et al.* 2009). However, scholars have been told that when compared with practising DL, improvement actions derived from practising SL are normally symptomatic treatments and would hardly move forward to perpetual performance improvement (Wong *et al.* 2009). To cope with an evolving business environment, contractors should generate new ways of working. In this regard, practising DL is more desirable.

It is worth noting that the cusp model is more suitable for use in describing the relationship among the practice of double-loop learning (DL), the attention to feedback (AF) and the anticipation of project performance (PP). It is because while fitting the data of Variables Sets 3, 4, 7,8, 11 and 12 to the cusp model, comparatively low loglikelihood, AIC and BIC values were found (see Table 7.5). Looking further into these variable sets, it is found that practice of 'DL1: identifying the root of the problem before taking improvement action' and 'DL2: seeking and adopting new management and working approach through evaluation of current practice' were used as the dependent variables. It is noted that these findings match with those previous studies that identified that learning is not a guaranteed action with the provision of project monitoring feedback (Love *et al.* 2000; Wong *et al.* 2010).

The results of this study reveal that performance improvement is merely one of the possible contractors' learning responses to project monitoring feedback. Sceptics have long argued that performance improvement may not be the ultimate learning response to project monitoring feedback (Love *et al.* 2000; Wong *et al.* 2010). Unless vulnerable to potential threats on losing prestige and future job opportunities, contractors rarely learn from feedback amicably and continuously (van Marrewijk 2007; Wong *et al.* 2010). In this sense, the traditional linear model that assumes positive relationships among practice of OL, AF and PP may sometimes be impractical to depict the contractor's learning response to the performance feedback. In this study, the fitness of the data set to the cusp model reveals that the practice of DL is conditional and is subjected to a sudden jump in view of anticipated performance. While contractors' attention to feedback may increase when performance depreciates, contracting organisations may remain not practising DL when the perceived performance still surpasses aspiration. A shift from not taking improvement action to taking improvement action occurs instantaneously when they perceive that performance is worse than expected.

As double-loop learning aims to devise new approaches of working by conducting a comprehensive review of the cause of failure, it has been identified as a more effective organisational learning practice to facilitate performance (Jashapara 2003). While researchers argued that contracting organisations in construction fail to practise double-loop learning, few of them had pinpointed the possible reasons (Chan *et al.* 2005; Wong *et al.*

2009). The results of this study indicate that anticipated performance can be understood as splitting factors of the practise of DL. More research is needed to examine the effect of these splitting factors on the practise of DL. This study adds new insights to the practice of different types of OL.

It has been advocated that the effect of OL on performance improvement does not come by chance (Kululanga *et al.* 1999). Findings in this study suggest that the persistent attention to feedback may be dependent on the contractor's aspiration of performance (Harrison and Palletier 1997; Simon 1955). When deciding to take improvement action in response to feedback, those who believe their performance is worse than expected would engage in DL in response to feedback. Inversely, those who believe their performance would be better than the pre-determined standards may opt for 'satisficing' (i.e. stop DL until performance is found worse than expectation) (Harrison and Palletier 1997; Simon 1955).

As such, the above results partly confirm the hypothesis H1, in which the contractors' practice of DL in response to their attention to feedback (AF) is contingent on their anticipation of project performance (PP) in the construction projects.

Summary

Research findings from previous studies indicate that learning behaviour upon feedback derived from project monitoring can be obscure. Following this stream of research, a dynamic model of organisational learning is developed in this study using the catastrophe theory modelling approach. Anticipation in meeting the pre-determined project goals represents the splitting factor of the proposed model, which suggests that learning from feedback may be conditional. The current study represents an endeavour to model the dynamic characteristic of learning behaviour in response to feedback. Construction contractors are mindful of the threats of losing prestige. Learning in these articles was delineated as a remedial action against the undesirable performance outcomes and consequences such as losing face. Nevertheless, while anticipating that their performance would exceed the pre-determined standards set by the clients, contractors may adopt a wait-and-see attitude towards feedback. Double-loop learning may happen when the contractors expect that their performance may be worse than the client's anticipation.

Researchers have placed much emphasis on feedback as a viable source of learning for contracting organisations' performance improvement (Huemer and Ostergsen 2000; Orange *et al.* 2005). Nevertheless, sceptics argued that unless vulnerable to negative feedback, contractors typically cease their learning response when the performance meets the client's expectation (Love and Josephson 2004; van Marrewijk 2007). The findings of this study may help to explain why such a paradox exists. While the contractor's attention to feedback may trigger the practice of SL, his practice of DL in response to feedback may be dichotomised (by their anticipation of project performance).

There is no intention to interpret the 'contractor's attention to feedback' and their 'anticipation of performance' as the only factors that affect the OL behaviour of the construction organisations. Lau and Rowlinson (2011) argued that the practice of opportunism may suppress the contractors to learn genuinely from the project monitoring feedback. Wong and Lam (2012) advocated that effective OL should be succeeded by 'unlearning'. They further operationalised unlearning as a removal of obsolete beliefs and routines by an organisation Wong and Lam (2012).

These are also relevant variables that may affect OL behaviour of the construction organisations. The major contribution of this study is to offer a new approach to comprehend the OL behaviours of the contractors. The findings affirm the appropriateness of employing attention to feedback and anticipation of project performance as the factors that may trigger discontinuous practice of DL. This provides a possible reason to explain the diverse findings reported in previous studies regarding how project monitoring feedback can trigger OL. This study adds new insights to the learning behaviour of the construction contractors. The results reveal that the practise of DL in response to feedback may suddenly cease when the contractors perceived that their performance has met the expectation. Contractors may have thresholds of a performance in their mind.

Acknowledgements

Special thanks to Mr. Kam Shing Wong and Miss Ka Yan Leung for collecting data for the study. The content of this chapter has been published in Volume 32(4) of the *International Journal of Project Management* and Volume 24(3) of the *Journal of Management in Engineering*. These are used with the permission from Elsevier and ASCE respectively.

References

Al-JiBouri, S. H. (2003) Monitoring systems and their effectiveness for project cost control in construction. *International Journal of Project Management*, 21(3), 145–54.

Brown, A. and Adams, J. (2000) Measuring the effect of project management on construction outputs: a new approach. *International Journal of Project Management*, 18(5), 327–35.

Chaaya, M. and Jaafari A. (2001) Cognizance of visual design management in life-cycle project management. *Journal of Management in Engineering*, 17(1), 49–57.

Chan, P., Cooper, R. and Tzortzopoulos, P. (2005) Organizational learning: conceptual challenges from a project perspective. *Construction Management and Economics*, 23(7), 747–56.

Chan S. L. and Leung N. N. (2004) Prototype web-based construction project management system. *Journal of Construction Engineering and Management*, 130(6), 935–43.

Cheung S. O., Suen C. H. H. and Cheung K. W. K. (2004) PPMS: a web-based construction project performance monitoring system. *Automation in Construction*, 13(3), 361–76.

Cheung, S. O., Yiu, T. W., Leung, A. Y. T. and Chiu, O. K. (2008) Catastrophic transitions of construction contracting behavior. *Journal of Construction Engineering and Management*, 134(12), 942–52.

Chow, P. T., Cheung, S. O. and Yiu, T. W. (2012) A cusp catastrophe model of withdrawal in construction project dispute negotiation. *Automation in Construction*, 22, 597–604.

Cobb, L. (1980) Estimation theory parameter estimation for the cusp catastrophe model, in *Proceedings of The Section on Survey Research Methods*. American Statistical Association, Washington, DC.

Cobb, L. and Zacks, S. (1985) Applications of catastrophe theory for statistical modeling in the biosciences. *Journal of the American Statistical Association*, 80(392), 793–802.

Crawford, P. and Bryne, P. (2003) Project monitoring and evaluation: a method for enhancing the efficiency and effectiveness of aid project implementation. *International Journal of Project Management*, 21(5), 363–73.

Dainty, A. R. J., Briscoe, G. H. and Millet, S. J. (2001) New perspectives on construction supply chain integration. *Supply Chain Management – An International Journal*, 6(4), 163–73.

Dikmen, I., Birgonul, M. T. and Ataoglu, T. (2005) Empirical investigation of organizational learning ability as a performance driver in construction, in A. Samad (ed.), *Knowledge Management in the Construction Industry: A Socio-Technical Perspective*. London: Ideas Group Publishing, 130–49.

Dou, W. and Ghose, S. (2006) A dynamic nonlinear model of online retail competition using cusp catastrophe theory. *Journal of Business Research*, 96(7), 838–48.

Everett, J. G. and Farghal, S. H. (1997) Data presentation for predicting performance with learning curves. *Journal of Construction Engineering and Management*, 123(1), 46–52.

Franco, L. A., Cushman, M. and Rosenhead, J. (2004) Project review and learning in the construction industry: embedding a problem structuring method within a partnership context. *European Journal of Operational Research*, 152(3), 586–601.

Garud, R. and van de Ven, A.H. (1992) An empirical evaluation of internal corporate venturing process. *Strategic Management Journal*, 13(5), 93–109.

Gilmore, R. (1981) *Structural Stability and Morphogenesis*. New York: Wiley.

Grasman, R. P. P. P., van der Maas, H. L. J. and Wagenmakers E-J. (2009) Fitting the cusp catastrophe in R: a cusp package primer. *Journal of Statistical Software*, 32(8), 1–27.

Greve, H. R. (2002) Sticky aspirations: organizational time perspective and competitiveness. *Organization Science*, 13(1), 1–17.

Harrison, E. F. and Pelletier, M. A. (1995) A paradigm for strategic decision success. *Management Decision*, 33(7), 53–59.

Hartelman, P. A. O. (1997) *Stochastic Catastrophe Theory*. Amsterdam: Faculteit der Psychologies.

Herbig, P. A. (1991) A cusp catastrophe model of the adoption of an industrial innovation. *Journal of Product Innovation Management*, 8(1), 127–37.

Huemer, L. and Östergren, K. (2000) Strategic change and organizational learning in two Swedish construction firms. *Construction Management and Economics*, 18(6), 635–42.

Ibbs C. W., Wong C. K. and Kwak Y. H. (2001) Project change management system. *Journal of Management in Engineering*, 17(3), 159–65.

Jashapara, A. (2003) Cognition, culture and competition: an empirical test of the learning organization. *The Learning Organization*, 10(1), 31–50.

Kagioglou, M, Cooper R., and Aouad, G. (2001) Performance management in construction: a conceptual framework. *Construction Management and Economics*, 19(1), 85–95.

Kaplan, R. S. and Norton, D. P. (1992) The Balanced Scorecard: measures that drive performance. *Harvard Business Review*, 70(1), 71–79.

Kululanga, G. K., McCaffer, R., Price, A. D. F. and Edum-Fotwe, F. (1999) Learning mechanisms employed by construction contractors. *Journal of Construction Engineering and Management*, 125(4), 215–23.

Landin, A. and Nilsson C-H (2001) Do quality system really make a difference? *Building Research and Information*, 29(1), 12–20.

Lau, E. and Rowlinson, S. (2011) The implications of trust in relationships in managing construction projects. *International Journal of Managing Projects in Business*, 4(4), 633–59.

Love, P. E. D., Heng, L. I., Irani, Z. and Faniran, O. (2000) Total quality management and the learning organization: a dialogue for change in construction. *Construction Management and Economics*, 18(3), 321–31.

Love, P. E. D. and Josephson, P-E. (2004) Role of error-recovery process in projects. *Journal of Management in Engineering*, 20(2), 70–79.

Love, P. E. D., Huang, J. C., Edwards, D. J. and Irani, Z. (2004) Nurturing a learning organization in construction: A focus on strategic shift, organizational transformation, customer orientation and quality centred learning. *Construction Innovation*, 4, 113–26.

Martin, L. and Root, D. (2010) Emerging contractors in South Africa: interactions and learning. *Journal of Engineering, Design and Technology*, 8(1), 64–79.

Mohamed, S. (2003) Scorecard approach to benchmarking organizational safety culture in construction. *Journal of Construction Engineering and Management*, 129(1), 80–88.

Murray, P., and Chapman, R. (2003) From continuous improvement to organisational learning: developmental theory. *Learning Organization*, 10(5), 272–82.

Oliva, T. A., Peter, M. H. and Murthy, H. S. K. (1981) A preliminary empirical test of a cusp catastrophe model in the social sciences. *Behavioral Science*, 26(1), 153–62.

Orange G., Burke A., Colledge, B. and Onions P. (2005) Knowledge management: facilitating organisational learning within the construction industry, in A. Samad (ed.), *Knowledge Management in the Construction Industry: A Socio-Technical Perspective*. London: Ideas Group Publishing, 166–84.

Raijmakers, M. E. J., van Koten, S. and Molenaar, P. C. M. (1996) On the validity of simulating stagewise development by means of PDP networks: application of catastrophe analysis and an experimental test of rule-like network performance. *Cognitive Science*, 20, 101–36.

Sense, A. J. and Antoni, M. (2003) Exploring the politics of project learning. *International Journal of Project Management*, 21(7), 487–94.

Shih, H. M. and Tseng, M. M. (1996) Workflow technology-based monitoring and control for business process and project management. *International Journal of Project Management*, 14(6), 373–78.

Simon, H. A. (1955) *A Behavioral Model of Rational Choice*. New York: Harvard Press.

Stamovlais, D. and Tsaparlis, G. (2012) Applying catastrophe theory to an information-processing model of problem solving in science education. *Science Education*, 96(3), 392–410.

Thom, R. (1975) *Structural Stability and Morphogenesis*, Reading, MA: W. A. Benjamin.

Toor, S. and Ogunlana, S. O. (2010) Beyond the 'iron triangle': stakeholder perception of key performance indicators (KPIs) for large-scale public sector development projects. *International Journal of Project Management*, 28(3), 228–36.

van der Maas, H. L. J. and Molenaar, P. C. M. (1992) Stagewise cognitive development: an application of catastrophe theory. *Psychological Review*, 99(3), 395–417.

van der Maas, H. L. J., Kolstein, R. and van der Pligt, J. (2003) Sudden transitions in attitudes. *Sociological Methods and Research*, 32 (2), 125–52.

van Marrewijk, A. (2007) Managing project culture: the case of Environ Megaproject. *International Journal of Project Management*, 25(3), 290–99.

Wagenmakers, E. J., van der Maas, H. L. J. and Molenaar, P. C. M. (2004) Fitting the cusp catastrophe model. Available at: www.psycg.nwu.edu/~ej/Encyclopediacadastrophe.pdf

Wong, C. H. (2004) Contractor performance prediction model for the United Kingdom construction contractor: study of logistic regression approach. *Journal of Construction Engineering and Management*, 130(5), 691–98.

Wong, P. S. P. and Cheung, S. O. (2005) From monitoring to learning: a conceptual framework, in *Proceedings of The 21st Annual Conference of the Association of Researchers in Construction Management*, SOAS, London, 1037–51.

Wong, P. S. P. and Lam, K. Y. (2012) Facing turbulence: a driving force for construction organizations to regain the unlearning and learning traction. *Journal of Construction Engineering and Management*, 138(10), 1201–11.

Wong, P. S. P., Cheung, S.O. and Leung, M. K. Y. (2008) The moderating effect of organizational learning type on performance improvement. *Journal of Management in Engineering*, 24(3), 162–72.

Wong, P. S. P., Cheung, S. O. and Fan, K. L. (2009) Examining the relationship between organizational learning styles and project performance. *Journal of Construction Engineering and Management*, 135(6), 497–507.

Wong, P. S. P., Cheung, S. O. and Wu, R. T. H. (2010) Learning from project monitoring feedback: a case of optimizing behavior of contractors. *International Journal of Project Management*, 28(5), 469–81.

Wong, P. S. P., Cheung, S. O., Yiu, R. L. Y. and Hardie, M. (2012) The unlearning dimension of organizational learning in construction projects. *International Journal of Project Management*, 30(1), 94–104.

Xiao, H. and Proverbs, D. (2003) Factors influencing contractor performance: an international investigation. *Engineering, Construction and Architectural Management*, 10(5), 322–32.

Yiu, T. W. and Cheung, S. O. (2006) A catastrophe model of construction conflict behaviour. *Building and Environment*, 41(4), 438–47.

142 *Peter Shek Pui Wong*

Yiu, T. W., Cheung, S. O. and Cheung, C. H. (2007) Toward a typology of construction mediator tactics. *Building and Environment*, 42(6), 2344–59.
Zeeman, E. C. (1977) *Catastrophe Theory: Selected Papers 1972–1977*. Reading, MA: Addison-Wesley.

Section D

Minimising non-productive use of scarce resources

8 Behaviour transition in construction contracting

Tak Wing Yiu and Sai On Cheung

Introduction

Conflicts are ubiquitous in construction projects (Bramble and Cipollini 1995; Carsman 2000; Fenn *et al.* 1997). They are often the inevitable outcomes of interactions among team members having diverging goals, attitudes, values, interests or beliefs (De Bono 1991; Rahim *et al.* 2000; Rhys Jones 1994). The fact is that they need to maximise the benefits to their own organisations. With different goals and needs, the potential for conflict is high when they work as a team (Cherns and Bryant 1984; Newcombe 1996; Walker 1989). One of the important tasks of project management is to handle conflicts properly, otherwise they may become disputes (Hibberd and Newman 1999; Kumaraswamy 1998), the presence of which may lead to detrimental effects on the work progress and the relationships between the contractual parties (Harmon 2001; Ock and Han 2003). Ability to manage conflict and minimise dispute gives competitive edge to a construction contracting organisation. The voluminous publications on construction conflict are clear evidence of its significance (Gardiner and Simmons 1992; Harmon 2003; Kumaraswamy 1998; Ock and Han 2003). These studies have provided useful references for identifying the types and consequences of construction conflicts. In particular, construction conflicts due to human factors have also been highlighted (Gardiner and Simmons 1992; Pretorious and Taylor 1986; Kharbanda and Stallworthy 1990). Interviews and case studies were often employed to provide qualitative analyses. For example, Gardiner and Simmons (1992) examined the causes of dysfunctional conflicts in the light of people-centred mechanisms by conducting structured interviews of the participants of 19 construction projects. In addition, a case study was presented to demonstrate the outcomes of positive interactions when inter-organisational orientations and team-building exercises are implemented at the early stage of a project. Furthermore, Harmon (2003) found that use of partnership and mediation as an intervention process could prevent and resolve conflicts. Notwithstanding that these anecdotal studies have successfully identified the causes of conflicts in the construction industry (Burton 1990; Butler 1973; Fenn 1991; Fenn *et al.*

1997; Friedman *et al.* 2000; Gardiner and Simmons 1998; Harmon 2003; Hellard 1987), the use of quantitative analysis can help researchers to explore construction conflicts further. To this end, a quantitative study of construction conflicts has been completed (Yiu and Cheung 2006). This work described the dynamic changes in construction conflict behaviour based on catastrophe theory. Using the same interacting variables defined in the study of Yiu and Cheung (2006) – i.e. construction conflict level, tension level and behavioural flexibility – this study employs moderated multiple regression (MMR) to examine the relationships between these three variables.

Catastrophic change in contracting behaviour

In the work of Yiu and Cheung (2006), catastrophe theory was applied to the study of construction conflict behaviour. Catastrophe theory is a branch of dynamical system theory; it was devised by Rene Thom (1975) and further popularised by Zeeman (1974, 1976, 1977). The theory applies to systems that undergo sudden changes in behaviour in response to a gradual change of the variables. Previous applications of the theory can be widely found in a number of behavioural studies in the perception (Stewart and Peregoy 1983; Taeed *et al.* 1988), client psychology (Callahan and Sashin 1990), management (Herbig 1991; Oliva *et al.* 1995), finance (Ho and Saunders 1980) and social sciences (Holyst *et al.* 2000; Flay 1978). In construction, Yiu and Cheung (2006) have successfully applied this theory in studying conflict behaviour as influenced by the level of conflict, which in turn depends on the interacting variables of tension level and behavioural flexibility. Behavioural flexibility is defined as the ability to act appropriately to the circumstances (Zaccaro *et al.* 1991a, 1991b). It is a type of personality trait that demonstrates the adjustment in one's behaviour responsive to the surroundings. This is reported to have implications for the conflict level (Walton and Dutton 1969). In the study of Yiu and Cheung (2006), behavioural flexibility was incorporated in the study of construction conflicts through the use of the model of conflict handling styles suggested by Rahim 1983; Rahim and Bonoma 1979). This model adopted the five conflict handling style classifications of Blake and Mouton (1964), i.e. integrating, obliging, compromising, dominating and avoiding. Moreover, two dimensions were added to Rahim's model. The first dimension describes the degree to which an individual attempts to satisfy his own concern. The second dimension explains the degree to which an individual wants to satisfy the concerns of the others. With the use of this model, it has been suggested that flexible individuals adjust their own conflict resolution styles according to the situation so as to maximise prospective collaboration (Glaser and Glaser 1991; Yiu and Cheung 2006).

In a catastrophe model, the interacting variables are called the normal factor and splitting factor. (Baack and Cullen 1992). The normal factor is related to the dependent variable (i.e. construction conflict behaviour) in a consistent pattern. The splitting factor is the key variable as it is described as 'a moderator variable which specifies conditions under which the normal factor will affect the dependent variable in a continuous fashion, and other circumstances under which the normal factor will produce discontinuous changes in the dependent variable ... it is the splitting factor that determines the "breaking point" or threshold of change in the dependent variable' (Baack and Cullen 1992). In the study of Yiu and Cheung (2006), the hypotheses of tension level as the normal factor and behavioural flexibility as the splitting factor were first established. Based on the algorithm suggested by Cobb (1980), Cobb and Watson (1980), Cobb *et al.* (1983) and Thom (1975), the 'catastrophe' of construction conflict behaviour was demonstrated with a robust program called Cuspfit, which was developed by Hartelman (1997), van der Maas and Molenaar (1992) and van der Maas *et al.* (2003). With this program, the appropriateness of the interacting variables and the fitness of the model were tested by checking goodness-of-fit indices such as Akaike's Information Criterion (AIC) and the Bayesian Information Criterion (BIC) (Akaike 1974; Cobb 1980; Cobb and Zacks 1985; Hartelman 1997; Ploeger *et al.* 2002; van der Maas and Molenaar 1992; van der Maas *et al.* 2003; Schwarz 1978). Based on these statistics, a final catastrophe model of construction conflict behaviour was developed.

A three-variable framework of construction conflict

Yiu and Chueng (2006) found that the bimodal in the nature of construction conflict behaviour could be detected using the three-variable framework, i.e. if a point residing in the model reaches a certain level of tension at a given level of behavioural flexibility, a sudden jump in conflict behaviour occurs. The hypotheses of tension level as the normal factor and behavioural flexibility as the splitting factor were also confirmed statistically. The relationships between tension level, construction conflict level and behavioural flexibility is shown in Figure 8.1.

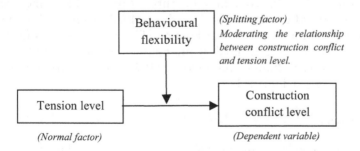

Figure 8.1 The three-variable framework of construction conflict (Yiu and Cheung 2006)

Technically, the three-variable framework can be described as a moderated causal relationship (Jaccard and Turrisi 2003), i.e. the relationship between tension level and construction conflict level is moderated by a third variable: behavioural flexibility. In other words, the nature of the relationship between tension level and construction conflict level varies, depending on the degree of behavioural flexibility. As suggested by Jaccard and Turrisi, such a relationship can be analysed with the use of moderated multiple regression (MMR). Under this method, tension levels are set as predictors, construction conflict levels as criteria and behavioural flexibility levels as moderator variables. The analysis involves examining the moderating effect of behavioural flexibility (B_i) on the relationship between tension level (T_i) and construction conflict level (C_i). If the moderating effect is not significant, it can then be said that the tension level has a 'constant' effect on the construction conflict behaviour (Cohen *et al.* 2003; Jaccard and Turrisi 2003). However, if such moderating effect was statistically significant, this would mean that a moderating effect is present in the relationship (Jaccard and Turrisi 2003). This study, which is built upon the earlier work of Yiu and Cheung (2006), specifically examines the application of MMR to the three-variable framework as shown in Figure 8.1. As the potential effect of behavioural flexibility is to modify the relationship between tension level and conflict level, this study enables project participants experiencing a high tension level to cope better with conflict situations in construction projects. Standard MMR procedures suggested by Jaccard *et al.* (1990) and Cohen *et al.* (2003) were used in this chapter.

The study

To study the relationships between construction conflict level, tension level and behavioural flexibility, a questionnaire was prepared to collect data for each of the three interacting variables. Each of these variables was measured using their sub-variables. For the construction conflict level, two sub-variables were used: adversarial attitudes and mistrust level (Arditi *et al.* 1998; Conlin *et al.* 1996; Harmon 2003; Mix 1997; Rhys Jones 1994; Steen and McPherson 2000). In comparison, three sub-variables, inconsistent demands, degree of uncertainty and work overloads, were used to measure the tension level (Burton 1990; Gardiner and Simmons 1998). As for behavioural flexibility, it was measured by integrating the conflict handling styles of Rahim (1983) and Rahim and Bonoma (1979). A summary of these sub-variables is given in Table 8.1.

Data collection

The survey was conducted in Hong Kong, and administered by post or faxed to the potential respondents who had been contacted and had expressed interest in participating. The targeted respondents were construction

Table 8.1 A summary of sub-variables

		Sub-variables and their measurement methods
The three-variable	Construction conflict level	• Adversarial attitudes among project teams* and; • Mistrust level among project teams* (Arditi *et al.* 1998; Conlin *et al.* 1996; Harmon 2003; Mix 1997; Rhys Jones 1994; Steen and McPherson 2000)
	Tension level	• Inconsistent demands from different project members* (Burton 1990; Gardiner and Simmons 1998); • Degree of uncertainty on the project* and; • Work overloads of project team members* (Gardiner and Simmons 1992, 1998).
	Behavioural flexibility	Details of the five-point scale (Rahim 1983a, 1983b; Rahim and Bonoma 1979; Blake and Mouton 1964): 1 High concern for both self and others (Integrating) — High Behavioural Flexibility 2 Low concern for self, high concern for others (Obliging) 3 Neutral (Compromising) 4 High concern for self, low concern for others (Dominating) 5 Low concern for self, low concern for others (Avoiding) — Low Behavioural Flexibility

Note: * A five-point Likert scale was adopted for each measurement (from 1 = least significant to 5 = most significant)

professionals such as project managers, engineers, architects and quantity surveyors who had experience in project management. The items included in the questionnaire for this survey are shown in Table 8.1. As the data needed to be case specific, the respondents were asked to select one of their most recent construction projects as a reference for the completion of the questionnaire.

The questionnaire has four sections. The respondents were requested to provide background information such as their professions and relevant experience in the first section. The next three sections address the measurement of sub-variables. The respondents were asked to rate the degree of significance of the construction conflict level and tension level on a Likert scale of 1 (least significant) to 5 (most significant). As discussed, the measurement of behavioural flexibility, the model of Rahim and Bonoma (1979), Rahim (1983) and Blake and Mouton (1964) was used and reduced to a Likert scale ranging from 1 (high concern for self and others) to 5 (low concern for self and others).

In this survey, a total of 200 questionnaires were sent out and 91 sets were completed and returned. This represents a response rate of 46 per cent. The returned questionnaires were completed by project managers (20 per cent), architects (20 per cent), engineers (19 per cent), quantity surveyors (38 per cent) and others such as construction lawyers and mediators (3 per cent). Most of the respondents were, at the time, holding senior positions in the industry, with 56 per cent having more than 10 years of experience. The composition of the respondents by their professions and working experience are shown in Tables 8.2 and 8.3.

The significant MMR models

As described previously, the tension level, construction conflict level and behavioural flexibility were used as predictors, criteria and the moderator variable respectively (Aiken and West 1991; Cobb and Zacks 1985; Darlington 1990; Jaccard *et al.* 1990). In this study, there were a total of six (231) moderated multiple regression models (devised from the combinations of sub-variables of the three-variable system as shown in Table 8.1). These models were then subjected to a test of significance of their interaction effects. The presence of a significant moderating effect is indicated if the inclusion of the predictor-moderator product (i.e. T_jB_k term) in the regression model produces a significant change in the R^2 (i.e. R^2) between equations (8.3) and (8.4). Based on the use of the F-test (Jaccard *et al.* 1990), two out of six MMR models were found to be statistically significant (see Table 8.4). However, significant moderating effects were not detected for the other four MMR models. In sum, the moderating effects

Table 8.2 Composition of respondents (by profession)

Professions	Number	Percentage
Project managers	18	20
Architects	35	38
Engineers	17	19
Quantity surveyors	18	20
Others (construction lawyers and construction mediators)	3	3
Total	91	100

Table 8.3 Composition of respondents (by working experience)

Working experience (years)	Number	Percentage
Below 5	23	25
5–10	17	19
11–15	10	11
16–20	17	19
Above 20	24	26
Total	91	100

Table 8.4 The six MMR models of construction conflict

| | Construction conflict level | | | | | | | |
| | Adversarial attitudes | | | | Mistrust level | | | |
	b_3	R^2	ΔR^2	test statistic[#]	b_3	R^2	ΔR^2	test statistic[#]
Tension level								
A Inconsistent demands × **Behavioural flexibility**	.018	.231	.000	.000	.007	.145	.000	.000
B Degree of uncertainty × **Behavioural flexibility**	–.085	.039	.005	.453**	–.099	.074	.006	.564**
C Work overloads × **Behavioural flexibility**	.046	.171	.002	.210	–.026	.091	.001	.100

Notes: [#]The test statistic was computed using Equation (5); p<0.10*, p<0.05** and p<0.01***

were significant in the relationships between: (1) the degree of uncertainty and behavioural flexibility on adversarial attitudes; and (2) the degree of uncertainty and behavioural flexibility on the mistrust level. These relationships are illustrated in Figure 8.2.

To discuss such findings further, the most commonly used method is to present the moderating effects with graphs. Regression lines were plotted for the regression of construction conflict level and tension level at the 'low' and 'high' values of behavioural flexibility. The 'low' value is defined as one standard deviation below the mean score and the 'high' value as one standard deviation above the mean (Jaccard and Turrisi 2003). This method has been successfully applied in similar studies using MMR (Etzion 1984; Jaccard and Turrisi 2003; Lim and Carnevale 1990; Pedhazur 1982). However, in this study, because the Likert scale for the measurement of behavioural flexibility ranged from 1 (high concern for self and for others)

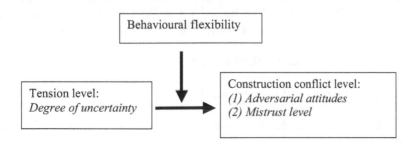

Figure 8.2 The two significant MMR models

to 5 (low concern for self and others), the meaning of 'low' and 'high' as defined by Jaccard and Turrisi (2003) required modification. The 'low' value of behavioural flexibility was defined as one standard deviation *above* the mean score, and the 'high' value of behavioural flexibility as one standard deviation *below* the mean. To generate these plots, the equation of MMR can be rewritten as:

$$
\begin{aligned}
C_i &= a + b_1T_j + b_2B_k + b_3T_jB_k + \varepsilon \\
&= (a + b_2B_k) + (b_1 + b_3B_k)T_j + \varepsilon \\
&= c + (b_1 + b_3B_k)T_j + \varepsilon
\end{aligned}
\tag{Eq. 8.1}
$$

$$
= c + b_{total}\,T_j + \varepsilon, \text{ where } c = (a + b_2B_k),\ b_{total} = b_1 + b_3B_k \tag{Eq. 8.2}
$$

where c is the intersection in the regression equation for the interaction model and b_{total} is the regression coefficient associated with the moderator variable in Equation 8.2.

According to the regression outputs, the two significant MMR models are:

$$
C_{AA} = 1.534 + .457T_{DU} + .364B + -.085T_{DU}B + \varepsilon \tag{Eq.8.3}
$$

$$
C_{ML} = 1.103 + .604T_{DU} + .390B + -.099T_{DU}B + \varepsilon \tag{Eq.8.4}
$$

where C_{AA} comprises the sub-variables of the construction conflict level related to adversarial attitudes. C_{ML} comprises the sub-variables of the construction conflict level related to the mistrust level. T_{DU} is the sub-variable of the tension level that relates to the degree of uncertainty.

Similarly, Equations 8.3 and 8.4 can also be written as:

$$
C_{AA} = (1.534 + .364B) + [.457 + (-.085)B]T_{DU} + \varepsilon \tag{Eq. 8.5}
$$

$$
C_{ML} = (1.103 + .390B) + [.604 + (-.099)B]T_{DU} + \varepsilon \tag{Eq. 8.6}
$$

For the variable of behavioural flexibility (B), the mean ($\bar{y}$) and standard deviation (SD) were found to be 3.23 and 0.94 respectively. Hence, the 'low' and 'high' degree of behavioural flexibility can be calculated.

$$
B_{low} = \bar{y} + SD = 3.23 + 0.94 = 4.17 \tag{Eq. 8.7}
$$

$$
B_{high} = \bar{y} - SD = 3.23 - 0.94 = 2.29 \tag{Eq. 8.8}
$$

where B_{low} and B_{high} are the 'low' and 'high' values of behavioural flexibility respectively.

Substituting the values of B_{low} and B_{high} into Equations 8.5 and 8.6, the following equations can be derived:

When the value of B_{low} in Equation (8.7) is substituted into Equation 8.5.

$$C_{AA} = (1.534 + .364B_{low}) + [.457 + (-.085)B_{low}]T_{DU} + \varepsilon$$
$$= 3.05 + .10T_{DU} + \varepsilon \qquad\qquad (Eq.\ 8.9)$$

When the value of B_{high} in Equation 8.8 is substituted into Equation 8.5.

$$C_{AA} = (1.534 + .364B_{high}) + [.457 + (-.085)B_{high}]T_{DU} + \varepsilon$$
$$= 2.37 + .26T_{DU} + \varepsilon \qquad\qquad (Eq.\ 8.10)$$

Similarly, when the value of B_{low} in Equation 8.7 is substituted into Equation 8.6.

$$C_{ML} = (1.103 + .390B_{low}) + [.604 + (-.099)B_{low}]T_{DU} + \varepsilon$$
$$= 2.73 + .19T_{DU} + \varepsilon \qquad\qquad (Eq.\ 8.11)$$

When the value of B_{high} in Equation 8.8 is substituted into Equation 8.6.

$$C_{ML} = (1.103 + .390B_{high}) + [.604 + (-.099)B_{high}]T_{DU} + \varepsilon$$
$$= 2.00 + .38T_{DU} + \varepsilon \qquad\qquad (Eq.\ 8.12)$$

From Equations 8.9 to 8.12, the regression lines for 'low' and 'high' degrees of behavioural flexibility for adversarial attitudes and mistrust level can be plotted. These are presented in Figures 8.3 and 8.4 respectively. If there is no interaction between the variables, the regression lines would all be parallel (Jaccard and Turrisi 2003; Pedhazur 1982). As the figures show that the lines of low behavioural flexibility are steeper than those of high behavioural flexibility, the interaction effects are indicated by the nonparallel lines in Figure 8.3 and 8.4.

As shown in Figures 8.3 and 8.4, the regression line for one group intersects with the corresponding regression line for the other group. To explain the meaning of these intersection points, the concepts of *ordinal* and *disordinal* interactions are applied. As suggested by Jaccard and Turrisi (2003), 'an ordinal interaction is one in which the regression lines are non-parallel but do not interact' and 'a disordinal interaction (also called a crossover interaction) is one in which regression line that regresses Y onto the continuous predictor for one group intersects with the corresponding regression line for the other group.' However, for any given set of non-parallel lines, there is always an intersection point. Theoretically, all interactions are disordinal in nature. Interactions are classified as ordinal if the regression lines do not intersect *within the range being studied* (Jaccard and Turrisi 2003; Pedhazur 1982). In Figures 8.3 and 8.4, as the intersection points appear within the range being studied, the plots reveal that the interaction is disordinal (or a crossover interaction).

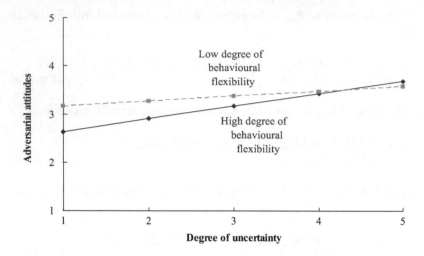

Figure 8.3 Moderating effect of behavioural flexibility on the relationship between degree of uncertainty and adversarial attitude

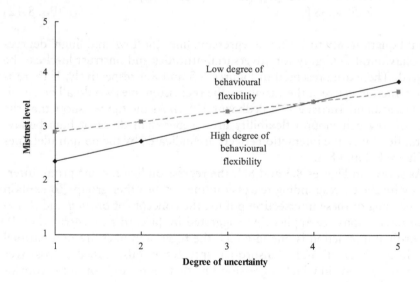

Figure 8.4 Moderating effect of behavioural flexibility on the relationship between degree of uncertainty and mistrust level

   To allow for a richer discussion, it is possible to identify the point where the regression lines intersect using Equation 8.13.

$$P = \frac{(a_1 - a_2)}{(b_2 - b_1)} \qquad \text{(Eq. 8.13)}$$

where P is the point of intersection (or crossover point), a_1, a_2 are the intercepts of the first and second regression lines respectively, and b_1 and b_2 are the slopes of the first and second regression lines respectively.

Hence, from Equation 8.9 to 8.12, the points of intersection are:

$$P_{AA} = (3.05 - 2.37) / (0.26 - 0.10) = 4.25 \qquad \text{(Eq. 8.14)}$$

$$P_{ML} = (2.73 - 2.00) / (0.38 - 0.19) = 3.84 \qquad \text{(Eq. 8.15)}$$

where P_{AA} and P_{ML} are the intersection points of the graphs in Figure 8.3 and 8.4 respectively.

These intersection points, P_{AA} and P_{ML}, obtained from Equations 8.14 and 8.15 describe the scores for the degree of uncertainty where the level of adversarial attitudes (or mistrust level) is the same under the low or high degree of behavioural flexibility. When the degree of uncertainty corresponds to a score of 4.25 (or 3.84), the level of adversarial attitudes (or mistrust) is predicted to be the same for low or high behavioural flexibility. If this score falls below 4.25 (or 3.84), the level of adversarial attitudes (or mistrust) is predicted to be higher under low behavioural flexibility than high behavioural flexibility. The intensification of adversarial attitudes and mistrust result from the behaviour of an inflexible individual who is unable to respond and adjust his own behaviour to the circumstances. On the other hand, a flexible individual is able to mitigate the level of adversarial attitudes (or mistrust). This is due to the fact that such an individual is willing to accommodate his own concerns in view of those of the others to avoid conflict escalation. However, as this score exceeds 4.25 (or 3.84), the level of adversarial attitudes (or mistrust) is predicted to be higher when the degree of behavioural flexibility is high. This implies that a flexible individual cannot mitigate the conflict level. Correspondingly, demarcation lines can be established in Figures 8.5 and 8.6 to demonstrate the effectiveness of behavioural flexibility. The behavioural flexibility is considered to be effective if it can mitigate the conflict level.

The left of the demarcation line is assigned to be the effective region of behavioural flexibility, while the ineffective region is assigned to the right of the line. As discussed by Cronbach and Glaser (1957) and Jaccard and Turrisi (2003), such treatments of disordinal interactions are commonly used by educational, organisational and psychological researchers. For example, decisions on the assignment of people to treatment, such as clinical interventions and types of educational curricula, are frequently guided by the identification of the intersection points in disordinal interactions. Those who fall to the left of the intersection point are assigned to the other

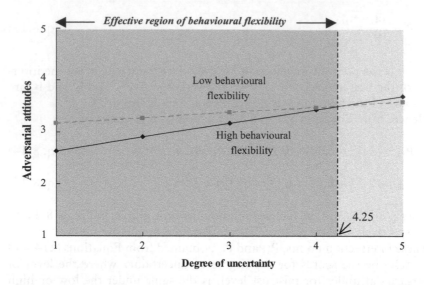

Figure 8.5 Effective region of behavioural flexibility for the relationship between degree of uncertainty and adversarial attitude

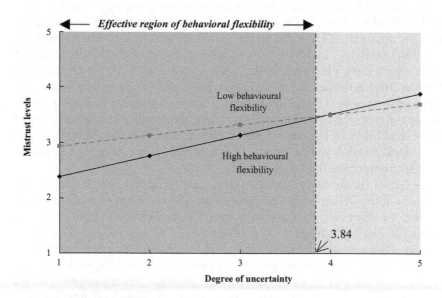

Figure 8.6 Effective region of behavioural flexibility for the relationship between degree of uncertainty and mistrust level

treatment (Jaccard and Turrisi 2003). In this study, the existence of disordinal interactions suggests a discontinuity in the effect of behavioural flexibility on tension–conflict relationships, i.e. the interaction with behavioural flexibility is not constant. Hence, if the degree of uncertainty reaches a threshold level of 4.25 or 3.84, the interaction between behavioural flexibility and tension–conflict relationships will change radically, i.e. even a flexible individual will not be able to minimise or resolve conflicts. This finding suggests that minimising uncertainty is the key factor in preventing such radical changes.

Summary

This study builds on the work of Yiu and Cheung (2006) and aims to examine the application of moderated multiple regression (MMR) to a three-variable framework of construction conflict. The findings suggest that not all MMR models display a significant moderating effect. The MMR models that were found to be significant affirm that the relationship between the degree of uncertainty and adversarial attitudes (or mistrust level) varies with the behavioural flexibility of the parties. With the aid of graphical representation, the interaction effects were found to be disordinal. The presence of disordinal interactions suggests that the interaction between behavioural flexibility and tension–conflict relationships can change radically. Such changes may be identified by the critical points of the degree of uncertainty. Beyond these points, a flexible individual may find it difficult to minimise or resolve construction conflicts. It is suggested that such changes could be prevented by minimising the degree of uncertainty in construction projects.

Acknowledgments

Part of the content of this chapter has been published in Volume 54(3) of the *IEEE Transactions on Engineering Management* and is used with the permission from IEEE.

References

Aiken, L. S. and West, S. G. (1991) *Multiple Regression: Testing and Interpreting Interactions*. Newbury Park, CA: Sage.

Akaike, H. (1974) A new look at statistical model identification. *IEEE Transactions on Automatic Control*, 19, 716–23.

Arditi, D., Oksay, F. E. and Tokdemir, O. B. (1998) Predicting the outcome of construction litigation using neutral networks. *Computer-Aided Civil and Infrastructure Engineering*, 13, 75–81.

Baack, D. and Cullen, J. B. (1992) A catastrophe theory model of technological and structural change. *The Journal of High Technology Management Research*, 3(1), 125–45.

Blake, R. R. and Mouton, J. S. (1964) *The Managerial Grid*. Houston: Gulf.

Bramble, B. B. and Cipollini, M. D. (1995) *Resolution of Disputes to Avoid Construction Claims*. Transportation Research Board, Synthesis of Highway Practice 24. Washington, DC: National Academy Press.

Burton, J. (1990) *Conflict: Resolution and Prevention*. Basingstoke: Macmillan.

Butler, A. G. (1973) Project management: a study in organizational conflict. *Academy of Management Journal*, 16, 84–101.

Callahan, J. and Sashin, J. I. (1990) Predictive models in psychoanalysis. *Behavioral Science*, 35, 60–76.

Carsman, H. W. (2000) Real-time dispute resolution, in *Proceedings of 2000 Mid-year Meeting: Real Time Dispute Resolution*. Associated General Contractors, Alexandria, Virginia.

Cherns, A. R. and Bryant, D. T. (1984) Studying the client's role in construction management. *Construction Management and Economics*, 2, 177–84.

Cobb, L. (1980) Estimation theory for the cusp catastrophe model, in *Proceedings of the Section on Survey Research Methods*. Washington, DC: American Statistical Association, 772–76.

Cobb, L. and Watson, B. (1980) Statistical catastrophe theory: an overview. *Mathematical Modeling*, 1, 311–17.

Cobb, L. and Zacks, S. (1985) Applications of catastrophe theory for statistical modeling in the biosciences. *Journal of the American Statistical Association*, 80, 793–802.

Cobb, L., Koppstein, P. and Chen, N. H. (1983) Estimation and moment recursion relations for multimodal distributions of the exponential family. *Journal of the American Statistical Association*, 78, 124–30.

Cohen, J., Cohen, P., West, S. G. and Aiken, L. S. (2003) *Applied Multiple Regression/Correlation Analysis for the Behavioral Sciences*. Mahwah, NJ: L. Erlbaum Associate.

Conlin, J. G., Langford, D. A. and Kennedy, P. (1996) The relationship between construction procurement strategies and construction contract disputes, in R. G. Taylor (ed.), *Proceedings of CIB W92 "North meet South" Procurement Systems Symposium*. Durban, South Africa, 66–82.

Cronbach, L. J. and Glaser, G. C. (1957) *Psychological Tests and Personal Decisions* (2nd edn). Urbana: University of Illinois Press.

Darlington, R. B. (1990) *Regression and Linear Models*. New York: McGraw-Hill.

De Bono, E. (1991) *Conflicts: A Better Way to Resolve Them*. London: Penguin Books.

Etzion, D. (1984) Moderating effect of social support on the stress-burnout relationship. *Journal of Applied Psychology*, 69(4), 615–22.

Fenn, P. (1991) Managing corporate conflict and resolution disputes on construction disputes, in *Proceedings of the 7th Annual Conference Association of Researchers in Construction Management (ARCOM)*. University of Bath, 22–33.

Fenn, P., Lowe, D. and Speck, C. (1997) Conflict and dispute in construction. *Construction Management and Economics*, 15(6), 513–18.

Flay, B. R. (1978) Catastrophe theory in social psychology: some applications to attitudes and social behaviour. *Behavioral Science*, 23, 335–50.

Friedman, R. A., Currall, S. C. and Tsai, J. C. (2000) What goes around comes around: the impact of personal conflict style on work conflict and stress. *The International Journal of Conflict Management*, 11(1), 32–55.

Gardiner, P. D. and Simmons, J. E. L. (1992) Analysis of conflict and change in construction projects. *Construction Management and Economics*, 10, 459–78.

Gardiner, P. D. and Simmons, J. E. L. (1998) Conflict in small and medium sized projects: case of partnering to the rescue. *Journal of Management in Engineering*, 14(1), 35–40.

Glaser, R. and Glaser, C. (1991) *Negotiating Style Profile*. King of Prussia, PA: Organization Design and Development.

Harmon, K. M. J. (2001) Pseudo arbitration clauses in New York City construction contracts. *Construction Briefings*, 7.

Harmon, K. M. (2003) Conflicts between owners and contractors: proposed intervention process. *Journal of Management in Engineering*, 19(3), 121–25.

Hartelman, P. A. I. (1997) *Stochastic Catastrophe Theory*. Amsterdam: Faculteit der Psychologie.

Hellard, R. B. (1987) *Managing Construction Conflict*. London: Longman Scientific Technical Press.

Herbig, P. A. (1991) A cusp catastrophe model of the adoption of an industrial innovation. *Journal of Product Innovation Management*, 8, 127–37.

Hibberd, P. and Newman, P. (1999) *ADR and Adjudication in Construction Dispute*, Malden, MA, Blackwell, Oxford.

Ho, T. and Saunders, A. A. (1980) A catastrophe model of bank failure. *The Journal of Finance*, 35(5), 1189–207.

Holyst, J. A., Kacperski, K. and Schweitzer, F. (2000) Phase transitions in social impact models of opinion formation. *Physica A: Statistical Mechanics and its Applications*, 285, 199–210.

Jaccard, J. and Turrisi, R. (2003) *Interaction Effects in Multiple Regression*. Thousand Oaks, CA: Sage.

Jaccard, J., Turrisi, R. and Choi, K. W. (1990) *Interaction Effects in Multiple Regression*. Thousand Oaks, CA: Sage.

Kharbanda, O. P. and Stallworthy, E. A. (1990) *Project Teams: the Human Factor*. Oxford: NCC Blackwell.

Kumaraswamy, M. M. (1998) Consequences of construction conflict: a Hong Kong perspective. *Journal of Management in Engineering*, 14(3), 66–74.

Lim, R. G. and Carnevale, P. J. D. (1990) Contingencies in the mediation of disputes. *Journal of Personality and Social Psychology*, 58(2), 259–72.

Mix, D. M. (1997) ADR in the construction industry: continuing the development of a more efficient dispute resolution mechanism. *Ohio State Journal on Dispute Resolution*, 12, 463.

Newcombe, R. (1996) Empowering the construction project team. *International Journal of Project Management*, 14(2), 75–80.

Ock, J. H. and Han, S. H. (2003) Lessons learned from rigid conflict resolution in an organization: construction conflict case study. *Journal of Management in Engineering*, 19(2), 89–89.

Oliva, T. A., Oliver, R. L. and Bearden, W. O. (1995) The relationships among consumer satisfaction, involvement, and product performance: a catastrophe theory application. *Behavioral Science*, 40, 104–32.

Pedhazur, E. J. (1982) *Multiple Regression in Behavioral Research: Explanation and Prediction*. New York: Holt, Rinehart, and Winston.

Ploeger, A., van der Maas, H. L. and Hartelman, P. A. (2002) Stochastic catastrophe analysis of switches in the perception of apparent motion. *Psychonomic Bulletin and Review*, 9(1), 26–42.

Pretorius, F. I. H. and Taylor, R. G. (1986) Conflict and individual coping behaviour in informal matrix organizations within the construction industry. *Construction Management and Economics*, 4, 87–104.

Rahim, M. A. (1983) A measure of styles of handling interpersonal conflict. *The Academy of Management Journal*, 26(2), 368–76.

Rahim, M. A., and Bonoma, T. V. (1979) Managing organizational conflict: a model for diagnosis and intervention. *Psychological Reports*, 44, 1323–44.

Rahim, M. A., Magner, N. R. and Shapiro, D. L. (2000) Do Justice perceptions influence styles of handling conflict with supervisors? What justice perceptions, precisely? *The International Journal of Conflict Management*, 11(1), 9–31.

Rhys Jones, S. (1994) How constructive is construction law? *Construction Law Journal*, 10(1), 28–38.

Schwarz, G. (1978) Estimating the dimension of a model. *Annals of Statistics*, 6, 461–64.

Steen, R. H. and McPherson, R. J. (2000) Resolving construction disputes out of court through ADR. *Journal of Property Management*, 65(5), 58.

Stewart, I. N. and Peregoy, P. L. (1983) Catastrophe theory modeling in psychology. *Psychological Bulletin*, 94, 336–62.

Taeed, L. K., Taeed, O. and Wright, J. E. (1988) Determinants involved in the perception of the Necker cube: an application of catastrophe theory. *Behavioral Science*, 33, 97–115.

Thom, R. (1975) *Structural Stability and Morphogenesis*. Reading, MA: W.A. Benjamin.

van der Mass, H. L. and Molenaar, P. C. M. (1992) Stagewise cognitive development: an application of catastrophe theory. *Psychological Review*, 99, 395–417.

van der Mass, H. L., Kolstein, R. and van der Pligt, J. (2003) Sudden transitions in attitudes. *Sociological Methods and Research*, 32, 125–52.

Walker, A. (1989) *Project Management in Construction* (2nd edn). Oxford: Blackwell Scientific.

Walton, R. E. and Dutton, J. M. (1969) The management of interdepartmental conflict: a model and review. *Administrative Science Quarterly*, 14(1), 73–84.

Yiu, T. W. and Cheung, S. O. (2006) A catastrophe model of construction conflict behaviour. *Building and Environment*, 41(4), 438–47.

Zaccaro, S. J., Foti, R. J. and Kenny, D. A. (1991a) Self-monitoring and trait-based variance in leadership: an investigation of leader flexibility across group situations. *Journal of Applied Psychology*, 76, 308–15.

Zaccaro, S. J., Gilbert, J. A., Thor, K. K. and Mumford, M. D. (1991b) Leadership and social intelligence: linking social perceptiveness and behavioral flexibility to leader effectiveness. *Leadership Quarterly*, 2, 317–42.

Zeeman, E. C. (1974) On the unstable behavior of stock exchanges. *Journal of Mathematical Economics*, 1, 39–49.

Zeeman, E. C. (1976) Catastrophe theory. *Scientific American*, 234(May), 65–83.

Zeeman, E. C. (1977) *Catastrophe Theory: Selected Papers 1972–1977*. Reading, MA: Addison-Wesley Publishing Company.

9 Equity in construction contracting negotiation

I: A study of behaviour–outcome relationship

Tak Wing Yiu

Introduction

Negotiation is seldom learned by construction practitioners as part of their formal education but rather through experience. Negotiation in the construction business involves a significant level of interaction among negotiators including project managers, engineers and surveyors. During the interaction process, negotiators attempt to reconcile their differences and reach mutual agreement by discussing their preferences (Mintu-Wimsatt and Calantone 1996). This process often involves face-to-face interaction and the exchange of information, concessions or compromise. These are known as the essential ingredients of effective negotiations (Graham *et al.* 1994; Nolan-Haley 1992). Face-to-face interaction plays an important role in the early stage of negotiation. It involves continuous communication that impresses each of the disputing parties. Such interaction influences future negotiation processes and reinforces the relationship between the negotiating parties (Graham *et al.* 1994). The exchange of information, concessions or compromise is often required for the needs and preferences of negotiators to be understood so that they can work in the same direction to obtain mutually beneficial outcomes (Graham *et al.* 1994; Pruitt and Kimmel 1977). Reciprocal interaction and exchange is fundamental to achieve negotiation success. It has been found that conducive negotiation interactions are characterised by reciprocal information exchanges (Putnam and Jones 1982): 'Integrative messages tend to be matched with integrative responses; while distributive communication tends to elicit distributive responses' (Goering 1997). Negotiators are presumed to find satisfaction in fair negotiations (ibid.) where input/outcome ratios are equal between the negotiating parties (Allen and White 2002). However, negotiators' perception of what is 'fair' can be subjective (Mintu-Wimsatt 2005). Some negotiators view certain elements as inputs, whereas others may view those same elements as outcomes (Tornow 1971), resulting in different perceptions of what constitutes equity and inequity (Foote and Harmon 2006). Negotiation in the construction industry is characterised by high degree of fragmentation, with numerous individual participants striving to meet their own goals and needs, and expecting to maximise their own benefits (Newcombe 1996; Walker 2002). Construction dispute negotiators should strive for fairness in interacting with their counterparts. As equity is at the heart of the concept of fairness (Messick and Sentis 1983), a better understanding of how negotiators (or their negotiating parties) respond to equitable or inequitable situations, which can either reinforce or undermine mutual respect and tolerance (Maoz 2005), would help in the creation of an equitable environment in construction dispute negotiation.

Equity sensitivity

Equity sensitivity extends the original equity theory and is more predictive and discriminant with regard to how individuals respond to feelings of inequity (Allen and White 2002). Equity sensitivity theory has been applied in research in business ethics (Kickul *et al.* 2005; Mudrack *et al.* 1999), job performance (Bing and Burroughs 2001), employee attitudes and behaviour (Kickul and Lester 2001; Shore *et al.* 2006), organisational citizenship behaviour (Blakely *et al.* 2005) and buyer-seller relationships (Mintu-Wimsatt 2005). Application of equity sensitivity in construction dispute negotiation is relatively limited. An overview of the relevant studies can be found in the work of King and Hinson (1994) and Allen and White (2002). Using case studies of hypothetical business situations, King and Hinson conducted laboratory experiments to examine the roles of equity sensitivity and sex in explaining negotiator relationships and cognitive orientations, and negotiation outcomes. Allen and White investigated equity sensitivity in business negotiation in under-award situations. However, these researchers (Allen and White 2002; King and Hinson 1994) used data from business students rather than from business practitioners. The reliability and relevance of the data may be challenged given the complexities and dynamics of actual business negotiation (Mintu-Wimsatt 2005). This study takes a more pragmatic approach and investigates equity sensitivity and negotiation using a sample of practitioners in the construction industry. The three objectives of this study are: (1) the development of an equity sensitivity construct of construction dispute negotiation to provide a framework to explain how negotiators in this field respond to equitable or inequitable situations; (2) the identification of generic types of negotiating behaviour and negotiation outcomes in the construction industry; and (3) the examination of the interrelationships among equity sensitivity, negotiating behaviour and negotiation outcomes. Research reveals that perceptions of equity affect negotiating behaviour (Mintu-Wimsatt 2005; Vecchio 1981). This study advances the present understanding of equity sensitivity and negotiation through the investigation of behaviour–outcome relationships. The findings of this study offer insight into the prediction of the reactions of construction negotiators in equitable or inequitable situations, provide practical information for the design of strategies for responding to inequitable situations and inspire further equity research in the area of construction dispute negotiation.

Adam's equity theory

The theoretical foundation of this study stems from Adam's equity theory (ET) (1963, 1965). This theory maintains that 'negotiators will behave consistently when faced with perceptions of inequity. When perceived inequity exists, negotiators will work to restore equity' (Mintu-Wimsatt 2005). This suggests that negotiators will share a universal preference that input/outcome ratios be equal among them (Foote and Harmon 2006). The core notion of ET is to model how individuals respond to under-reward situations in an attempt to bring their equity ratio back into balance and manage their relationships with others. Specifically, when perceived inequity exists, tension is created among individuals, which motivates them to restore equity. Inequity is defined as either an under-reward or over-reward situation. In an under-reward situation, an individual's input/outcome ratio is less than that of the counterpart; in an over-reward situation, the individual's input/outcome ratio is greater than that of the counterpart. Based on the assumption that individuals are equally sensitive to equity, individuals who perceive themselves as either under-rewarded or over-rewarded will feel distress. The greater the distress an individual feels, the harder he or she will work to restore equity. The aim of equity restoration is to bring equity ratios back into balance (Allen and White 2002; Huseman *et al.* 1987; Mintu-Wimsatt 2005; Sauley and Bedeian 2000). This corresponds to what has been termed the 'norm of equity' (Huseman *et al.* 1987), which has been identified in both laboratory studies and field research. However, ET has been criticised for its inability to predict exactly how individuals will respond to inequitable situations (Allen and White 2002; Greenberg 1990). Therefore, ET has been extended to include the element of individual differences. This is the notion of equity sensitivity, which is an individual difference variable that explains how individuals react to inequity (Huseman *et al.* 1987). Equity sensitivity is related to 'an individual's perception of what is and what is not equity and then uses that information to make predictions about reactions to inequity' (King *et al.* 1993, in Mintu-Wimsatt 2005). More recent developments in ET have been spurred by the development of the equity sensitivity construct in the fields of management and organisational behaviour (Hartman and Villere 1990; Huseman *et al.* 1985; Huseman *et al.* 1987; Kickul and Lester 2001; King *et al.* 1993; Shore *et al.* 2006). Huseman *et al.* (1987) and Mintu-Wimsatt (2005) propose that equity sensitivity can be conceptually understood by identifying the characteristics of individuals as points along a continuum (see Figure 9.1). On one end of this continuum are benevolents. These types of negotiators prefer their input/outcome ratio to be less than that of the

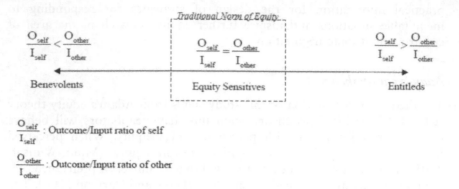

Figure 9.1 Equity sensitivity continuum (Huseman *et al.* 1987; Mintu-Wimsatt 2005)

counterparts. They are also known as 'givers' who are willing to provide inputs to their counterparts. With a high tolerance for being under-rewarded, they prefer giving to receiving and make a valuable contribution to the relationship. In addition, they express distress either when outcome/input ratios are equal or when their ratio is greater (Huseman *et al.* 1987; King *et al.* 1993; McLoughlin and Carr 1997; Miles *et al.* 1989; Mintu-Wimsatt 2005; Sauley and Bedeian 2000). Equity sensitives resides in the middle of the continuum. The notion of what constitutes equity for these types of negotiators comes close to the traditional norm of equity, as they desire balanced input/outcome ratios. They feel distress when under-rewarded and guilt when over-rewarded (Huseman *et al.* 1987; Mintu-Wimsatt 2005; Sauley and Bedeian 2000). On the other end of the continuum are the entitleds. These types of negotiators are known as 'takers' who focus mainly on their own outcomes. They often take action to achieve their objectives that results in imbalance in input/outcome ratios. They are not concerned about the outcomes of others and feel little or no obligation to reciprocate. They prefer getting to giving, and try to take more for themselves than they give to others (Allen and White 2002; Huseman *et al.* 1987; King *et al.* 1993; Miles *et al.* 1989; Mintu-Wimsatt 2005; Sauley and Bedeian 2000). Entitleds do not tolerate being under-rewarded, and are more tolerant of being over-rewarded than either equity sensitives or benevolents (King *et al.* 1993).

Research model and hypothesis

Negotiators distribute resources based on the contributions of their counterparts, and it is proposed that negotiation acts as equal reciprocal exchange (Goering, 1997; Leventhal 1976). However, equity sensitivity suggests that benevolent negotiators are givers who are willing to put in more effort than are their counterparts (King *et al.* 1993), whereas entitleds are takers who focus on themselves and their outcomes (Mintu-Wimsatt 2005), and thus are less likely to cooperate than are their counterparts. Equity sensitivity refers to a person's perception of equity, and hence influences the choice of negotiating strategy of negotiators (Thompson 1990). King and Hinson (1994) found that equity sensitivity affects how negotiators evaluate their interactions with their opponents. As negotiation strategies vary according to negotiating behaviour, which includes the response of negotiators to inequity, a relationship is hypothesised between equity sensitivity and negotiating behaviour during negotiations. If equity sensitivity and negotiating behaviour are related, then they will influence negotiation outcomes. This is supported by research that finds that individuals' perception of 'degree of equity' is a major determinant of job effort, performance and satisfaction (Huseman *et al.* 1987; Mueller and Clarke 1998). Sauley and Bedeian (2000) give clues to explain how equity sensitivity affects outcomes. They suggest that giving is an internally controlled outcome because individuals can decide how much they wish to give. Conversely, getting is an external controlled outcome because what one receives depends on what others are willing to give. Entitleds focus on what they can get from an exchange (externally controlled outcome), whereas benevolents are concerned with what they can give to an exchange (internally controlled outcome). In terms of outcome satisfaction, Huseman *et al.* (1987) found that benevolents express a high degree of satisfaction when under-rewarded, entitleds show a high degree of satisfaction when over-rewarded and equity sensitives prefer to be equitably rewarded.

In this study, the relationships between equity sensitivity, negotiating behaviour and negotiation outcomes are established by hypothesising negotiation outcomes as a dependent variable, negotiating behaviour as an independent variable and equity sensitivity as a moderating variable. The research model is graphically shown in Figure 9.2.

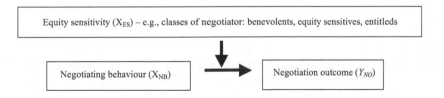

Figure 9.2 The research model

Methods

Measures

To achieve the aforementioned specific objectives, a questionnaire survey was conducted. Three types of data were included: those of (1) the 16-item modified Equity Preference Questionnaire (EPQ); (2) the reported use of negotiating behaviour; and (3) the achievement of negotiation outcomes.

(1) Equity sensitivity

Equity sensitivity was measured using the 16-item EPQ of Sauley and Bedeian (2000). This instrument is reliable and easy to administer. EPQ was designed to measure three groups of the equity sensitivity construct: benevolents, equity sensitives and entitleds (ibid.). It was designed on the basis of four types of studies:

(a) two pilot studies to purify the EPQ and to assess its reliability;
(b) two validity assessment studies to examine the validity of the EPQ;
(c) a laboratory experiment to determine the validity of the EPQ for predicting satisfaction with different reward conditions;
(d) a test-retest reliability study to provide evidence on the consistency of measurement yielded by the EPQ across time.

The attributes of the EPQ were modified to suit the context of construction dispute negotiation.

(2) Negotiating behaviour and negotiation outcomes

Types of negotiating behaviour and negotiation outcomes were measured with 14 items developed based on a literature search (see Tables 9.2 and 9.3).

The respondents were required to indicate the degree of their agreement with the listed types of negotiating behaviour on a Likert scale from 1 (strongly disagree) to 7 (strongly agree), and the degree to which they had achieved the listed negotiation outcomes on a scale from 1 (not achieved) to 7 (highly achieved).

Data collection

A questionnaire survey was used to collect data from negotiators experienced in construction disputes. According to the sample planning proposed by Luck and Rubin (1987), the first step is to define the target population(s) to be sampled. In this study, the target populations included project managers, architects and surveyors. One of their typical tasks is to resolve disputes or claims through negotiation. Respondents were asked to reflect on one of their most recently completed negotiation cases when completing the questionnaire. They were selected from a list of construction business firms from the Builder Directory and the Web pages of professional institutes including the Hong Kong Institute of Architects (HKIA) and the Hong Kong Institute of Surveyors (HKIS). Simple random sampling was used to draw a sample from each of the target populations. A similar approach has been successfully employed in the studies of Fang *et al.* (2004) and Fong and Chu (2006). The targeted respondents were contacted, and if they agreed to participate in the questionnaire survey, then the questionnaire was sent to them by post, fax or email, according to their preference.

Results

A total of 180 questionnaires were sent to the construction professionals of these target populations, and 83 respondents completed them, with a response rate of 46 per cent. More than 50 per cent of the respondents had more than five years' experience in construction dispute negotiation; 40 per cent were employees of surveying consultants, 35 per cent worked for main contractors, 20 per cent worked for private developers and 5 per cent worked for independent consultants. With the collected data, an equity sensitivity construct of construction dispute negotiation was developed using principal component factor analysis (PCFA). The suitability of the data was first assessed using the Kaiser-Meyer-Olkin (KMO) measure of sampling adequacy and Bartlett's test. The value of the KMO measure was 0.668, which is greater than the required threshold requirement of 0.5 (Cheung and Yeung 1998; Holt 1997). The low significance of Bartlett's test suggested the suitability of the data set for PCFA. The PCFA results gave a factor structure that represented the equity sensitivity construct of construction dispute negotiation (see Table 9.1).

Table 9.1 shows that the PCFA results did not yield exactly the same three-factor structure of the equity sensitivity construct (benevolents, equity sensitives and entitleds) suggested by Huseman *et al.* (1987) and Mintu-Wimsatt (2005); rather, a more detailed classification of entitleds was obtained as follows: Factor 1: Entitleds – unwillingness to reciprocate; Factor 4: Entitleds – self-interest; and Factor 5: Entitleds – press of outcome. To describe this five-factor solution, confirmatory factor analysis was performed using AMOS. Following Blakely *et al.* (2005), the confirmatory fit index (CFI) and root mean square error of approximation (RMSEA) were employed to assess the model fitness. The CFI value and RMSEA were 0.905 and 0.066, respectively. These collectively indicated that the data fit the model adequately.

Next, the constructs for negotiating behaviours and negotiation outcomes were identified. To explore their data structures, PCFA was performed to consolidate the results and facilitate interpretation. With the same criteria as those used previously to extract the number of factors, a

Table 9.1 Equity sensitivity construct of construction dispute negotiation

Equity sensitivity construct	Factors				
	1	2	3	4	5
Factor 1: Entitleds - unwillingness to reciprocate					
I try to get out of the negotiation table (Q3).	.800	−.225	−.092	−.101	.720
If I can leave the negotiation table, I try to work just a little more slowly than my counterparts expect (Q4).	.738	−.129	.010	.198	.602
I prefer to do as little as possible while getting as much as I can from my counterparts (Q1).	.728	−.126	−.304	.114	.722
I am most satisfied when I expend as little effort as possible (Q2).	.707	−.025	.131	.117	.572
Factor 2: Benevolents					
I feel obligated to negotiate the best deal for my client (Q11).	.072	.777	.076	−.024	.686
When I have completed my task, I help out other negotiation partners who have yet to complete their tasks (Q8).	−.194	.743	−.064	.115	.641
Even if I reach a non-desirable negotiation outcome, I will still try to do my best to settle the dispute/claim (Q9).	−.251	.675	.300	.050	.613
I prefer to handle many issues rather than waste time (Q13).	−.234	.513	.098	.149	.609
Factor 3: Equity sensitives					
I feel uneasy when there is little work for me to do (Q14).	−.102	.123	.847	−.050	.794
I become very dissatisfied if I handle few or no issues (Q15).	.150	−.008	.804	.098	.778
If the duties are equal among my negotiation partners, then it is better to deal with complex issues rather than a few issues (Q16).	−.213	.249	.654	.392	.756
Factor 4: Entitleds – self-interest					
A wise negotiator is concerned about his own outcomes rather than his inputs (Q7).	.046	.071	.046	.794	.698
It is a smart negotiator who gets as much as he can while giving as little as possible in return (Q5).	.064	−.024	.067	.744	.593
It is really satisfying when I can gain something for nothing (Q6).	.441	.205	.064	.646	.660
Factor 5: Entitleds – press of outcome					
If the negotiation has gone on for a long time, I will probably quit (Q10).	.069	.014	−.026	.099	.809
My greatest concern is whether or not I can achieve a desirable negotiation outcome (Q12).	−.222	.252	.179	.192	.386

four-factor solution was developed for both negotiating behaviours and negotiation outcomes. The PCFA results also satisfied the statistical fitness criteria of the KMO measure and Barlett's test. The KMO values for the PCFA of negotiating behaviours and negotiation outcomes were 0.723 and 0.772 respectively, and both obtained low significance in Bartlett's test, which suggested the suitability of the data set for PCFA. Hair *et al.* (1998) suggested that a factor loading value of 0.60 is a good demarcation for variable selection within factors. Hence, variables with a loading of less than 0.60 were discarded to achieve a simpler structure with greater interpretability (Fava and Velicer 1992). The final factor structures of negotiating behaviour and negotiation outcomes are given in Tables 9.2 and 9.3, respectively.

Negotiators in the construction business adopt different types of negotiating behaviour to maximise their benefits. The results of factor analysis revealed the solution-focused approach (Factor 1), aggressive approach (Factor 2), cooperative approach (Factor 3) and dominating approach (Factor 4) to be the major types of negotiating behaviour used by negotiators in construction disputes. These negotiators apparently adopt pragmatic approaches to resolve negotiation issues, such as the solution-focused approach (Factor 1). This approach aims to create a collaborative rather than a competitive climate, focusing on the problem and getting both negotiating parties to work out a solution. It is an approach that can help to achieve a satisfactory solution for both negotiating parties (Hodgson 1996). In contrast, aggressive negotiators (Factor 2) may use distributive strategies to change the attitudes, attributions or actions of their counterparts; to induce concessions from their counterparts (Walton and McKersie 1965; Baron and Richardson 1994) and often seek to achieve their goals at the expense of their counterparts (Graham *et al.* 1994; Rubin and Brown 1975; Rinehart and Page 1992; Monge *et al.* 1997; Nolan-Haley 1992). An aggressive approach has a negative effect on the negotiation relationship, in limiting the information exchange (Olekalns *et al.* 1996). A bad impression is thus given, which affects future relationship (Frazier and Summers 1984). Cheating and using threats are probably the most "effective" ways of escalating a dispute to a deadlock (Hodgson 1996). In construction dispute negotiation, the power of the developer (or developer's representative) and contractor is often uneven. Threats are often used by a developer as a tool to express his power. However, the effectiveness of a threat depends on the credibility of the threatening party's intention to carry it out and the amount of damage (e.g., in an on-going business relationship) that it could cause (Hodgson 1996). A cooperative approach (Factor 3) can be adopted to achieve mutually satisfying outcome. Such an approach involves a high level of information exchange, concession making, joint conflict resolution and integration (Bazerman *et al.* 2000; Graham *et al.* 1994; Pruitt 1981; Rubin and Brown 1975). Such interactions help in the maintenance of a positive relationship, and facilitate the achievement of mutual outcomes

Table 9.2 Factor structure of types of negotiating behaviour

Factor structure of types of negotiating behaviour	Factors			
	1	2	3	4
Factor 1: Solution-focused approach				
If problems arise, then I am willing to ask questions.[1]	.826	−.121	.015	.222
When problems arise, I am willing to solve the dispute with my counterparts. [3]	.733	−.029	.347	−.027
I expect to achieve outcomes that are beneficial to my company. [1-2]	.707	.105	.202	.251
*At work, I am willing to assist my counterparts to solve problems. [1]	.497	−.473	.085	.293
*It is essential to build a trustful working environment. [4]	.477	−.053	.320	−.283
Factor 2: Aggressive approach				
I will cheat and threaten my counterparts if necessary. [1-2] [6]	.003	.799	.122	−.163
I will make excessive demands of my counterparts. [1-2] [6]	−.163	.701	−.084	.219
I will achieve my own goal only at the expense of the other parties. [1-2]	−.033	.653	−.405	.151
*My greatest concern is to ensure my relatively favourable individual settlement. [1-2]	.355	.493	.157	.247
Factor 3: Cooperative approach				
I like to cooperate with my counterparts. [1-2] [6]	.213	−.097	.774	.025
Once my counterparts' requirements are fully understood, I will try my best to satisfy their needs. [1]	.059	.041	.694	.337
I will discuss with my counterparts both of our needs and preferences. [1]	.387	−.001	.606	.110
Factor 4: Dominating approach				
I will change my counterparts' attitudes, attributions or actions to achieve my expected goal. [1] [5]	.095	.166	.093	.796
I will try to induce concessions from my counterparts. [1] [5]	.330	−.052	.359	.616

Note: *Discarded item – factor loading < 0.60.

References:
(1) Graham *et al.* 1994
(2) Monge *et al.* 1997
(3) Sherif *et al.* 1965
(4) Luo 2002
(5) Mintu-Wimsatt and Calantone 1996
(6) Crane *et al.* 1999

Table 9.3 Factor structure of negotiation outcomes

Factor structure of negotiation outcomes	Factors			
	1	2	3	4
Factor 1: Integrative agreement				
Concessions were made by the negotiation partners. [5]	.753	−.047	.133	.166
The relationship between parties was harmonious and the possibility of dealing with each other in the future increased. [3]	.680	.158	.283	−.121
Trust developed between parties. [1] [2]	.675	.313	.204	−.270
Communication between parties increased. [7]	.625	.304	.431	−.195
*Both parties' expectations were met. [4]	.526	.363	−.068	−.360
*My counterparts' strategy was adopted. [13]	.400	.310	.035	.393
Factor 2: Time saving				
Future disagreements are less likely. [6]	.146	.876	.137	−.139
The time required to solve problems was reduced. [11]	.085	.751	.332	.145
*The level of conflict was reduced. [6]	.381	.529	.151	−.233
Factor 3: Effective information transmission				
Information exchange increased. [8]	.227	.143	.780	−.142
A mutually beneficial solution was created. [9]	.167	.175	.767	−.066
*Organisational decision making improved. [10]	.173	.511	.573	.013
Factor 4: Deterioration of relationship				
A climate of hostility and distrust developed. [10]	−.310	−.115	.037	.772
My counterparts' needs were not clearly defined. [6]	.062	−.032	−.358	.724

Note: *Discarded item – factor loading < 0.60.

References:
(1) Plowman 1998
(2) Dozzi *et al.* 1996
(3) Schawarz and Peutsch 2001
(4) Luo 2002
(5) Graham *et al.* 1994
(6) Friedman *et al.* 2000
(7) Turner 2004
(8) Gulati 1995
(9) Bennett & Jayes 1995
(10) Rahim 2002
(11) Sheppard *et al.* 1989
(12) Porter 1979
(13) Mintu-Wimsatt and Calantone 1996

(Dozzi *et al.* 1996; Graham *et al.* 1994; Olekalns *et al.* 1996; Monge *et al.* 1997). This reciprocal approach is common in negotiations. Negotiators have to adjust their strategies or to make concessions in response to the actions of their negotiation counterparts. For instance, in pleasant conditions, negotiators expect more favourable negotiation outcomes and make more concessions (Baron 1990). If one of the negotiating parties uses a cooperative approach, then the others are more likely to respond in the same manner. Conversely, if a negotiator appears to be aggressive, then his or her negotiation partners will respond aggressively. Luo (2002) indicated that cooperation is initiated because of the negotiator's desire to ensure future social exchanges and maintain an on-going business relationship. In the construction industry, there are a relatively small number of large local contractors and large number of small or medium local contractors. Therefore, the relationships among practitioners are often complementary, as larger contractors often subcontract work to small or medium contractors. The development of harmonious working relationships is absolutely essential to survive in the construction business. In the context of negotiation, it is thus important to establish a cooperative relationship at the early stage of negotiation: future reciprocity can then be reinforced and the cooperative approach be sustained (Dabholkar *et al.* 1994; John and Jack 2005; Mintu-Wimsatt and Calantone 1996; Monge *et al.* 1997). Finally, dominating approach (Factor 4) used by negotiators of construction disputes emphasises concern for self and undermines the concern of others (Rahim *et al.* 2000). Dominating negotiators go all out to achieve their goals by changing the attitudes, attributions or actions of their counterparts. In construction, this approach may be appropriate when a speedy decision is required or a routine matter is involved.

Four generic types of negotiation outcomes were classified by PCFA: satisfaction (Factor 1); time saving (Factor 2); effective information transmission (Factor 3); and deterioration of relationship (Factor 4). Among these, the first three negotiation outcomes are considered functional, and the last is considered dysfunctional. Functional outcomes yield positive effects, including making concessions, building trust, fostering communication and saving time. Dysfunctional outcomes have negative effects such as the creation of hostility and distrust during the negotiation process. Factor 1 collectively describes satisfaction, which is delimited as satisfaction with successfully making concessions, improving relationships, building trust and facilitating communication. This negotiation outcome can be achieved if a dispute can be settled with a solution that satisfies both negotiating parties. In construction dispute negotiation, this implies that both parties motivate each other to provide better value by aligning each organisational objective with project objectives. Factor 2 is described as time saving. This factor refers to the efforts of each negotiation party to reduce future disagreements and increase the efficiency of negotiation. If a dispute becomes inevitable, then it is important to manage it positively to encourage early and effective settlement. Otherwise, the dispute will need to be resolved by expensive and time-consuming arbitration or litigation. Factor 3 is related to effective information transmission in construction dispute negotiation. Information is the set of data, facts or opinions that is directly related to the conducting of negotiations. This information can be trans-mitted through different communication channels such as face-to-face meetings or formal correspondence. Communication is an essential negoti-ation instrument, and negotiation is impossible without each side understanding the other side's concerns (Nieuwmeijer 1988). Thus, effective information transmission can foster mutually beneficial solutions. Finally, a dysfunctional outcome, the deterioration of the relationship (Factor 4), was reported. Working relationships can deteriorate if serious problems arise such as claims for damages or other monetary issues. Relationships are likely to break down if both negotiating parties seek to maximise their own benefits without establishing an effective process for resolving important differences. This outcome often results from confrontational contracting behaviour. It is well known that relationships are a prime factor in business dealings. This is particularly so in the construction industry, because devel-opers will not invite contractors with whom they have bad relationships to tender for their projects. With skilful negotiation, ongoing business devel-opments can progress and working relationships can be sustained.

Moderating effects of equity sensitivity on the behaviour–outcome relationship

Based on the five-factor solution of the equity sensitivity construct of construction dispute negotiation and the four-factor solutions of negotiat-ing behaviours and negotiation outcomes, composite scales were devised. These scales are the composite measures created for each observation on each factor extracted via PCFA (Hair *et al.* 1998). New sets of variables were thus calculated for moderated multiple regression (MMR).

Next, the research model depicted in Figure 9.2 was statistically tested using MMR. This is a common statistical technique that is used to quantify the relationships between two or more predictor variables and dependent variables (Berry and Feldman 1985; Cobb *et al.* 1983; Lewis-Beck 1980). Specifically, this research model can be interpreted as follows: the predictive power of negotiating behaviour (X_{NB}) for negotiation outcome (Y_{NO}) depends on equity sensitivity (X_{ES}). The significance of the moderating vari-able, X_{ES}, was tested using MMR. A hypothetical MMR model was constructed as follows:

$$Y_{NO_i} = a_0 + b_1 X_{NB_j} + b_2 X_{ES_k} + b_3 X_{NB_j} X_{ES_k} + \varepsilon \qquad \text{(Eq. 9.1)}$$

where Y_{NO_i}, X_{NB_j} and X_{ES_k} are the composite scales of negotiation outcomes (where i = 1, 2, 3 or 4), types of negotiating behaviour (where j = 1, 2, 3 or 4) and equity sensitivity (where k = 1, 2, 3, 4 or 5), respectively. $X_{NB_j} X_{ES_k}$ is the moderating term.

To test the research model, the MMR procedure suggested by Jaccard *et al.* (1990) and Cohen *et al.* (2003) was adopted. This moderating effect can be examined if a moderating term, $X_{NB}X_{ES}$, is included in Equation 9.1. In this study, a total of eighty (5 × 4 × 4) MMR models were developed (devised from the combinations of the five items of the composite scale of the equity sensitivity construct, four items of the composite scale of negotiating behaviour and four items of the composite scale of negotiation outcomes). For each of these MMR models, the significance of the moderating effect was then tested based on the MMR procedure of Jaccard *et al.* (1990) and Cohen *et al.* (2003). Finally, six MMR models were found to be significant. A similar approach was successfully employed by Lim and Carnevale (1990), who examined ninety MMR models based on composite scales. The six significant models provide evidence that the equity sensitivity of negotiators significantly moderates the behaviour–outcome relationship. To facilitate the interpretation of this finding, all of these models were combined to form an overall framework for further discussion (see Figure 9.3).

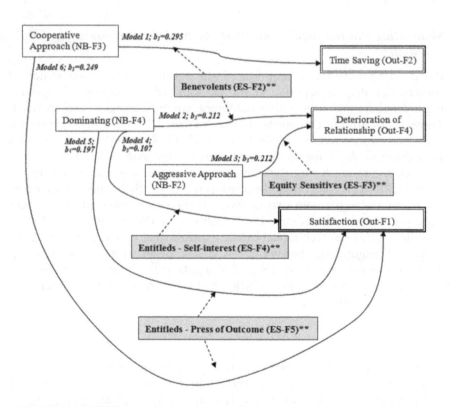

Figure 9.3 Framework of the moderating effects of equity sensitivity on behaviour–outcome relationships (significance of the moderating effects: p < 0.05**)

The results of MMR suggest that equity sensitivity plays a significant role in moderating the behaviour–outcome relationship in construction dispute negotiation. We structure the discussion of Figure 9.3 based on the three major types of negotiators of construction disputes: (1) benevolent negotiators; (2) equity sensitive negotiators; and (3) entitled negotiators.

Benevolent negotiators

Benevolent negotiators have a greater tolerance for being under-rewarded and experience guilt when they perceive that they are equitably rewarded or over-rewarded (Foote and Harmon 2006). Models 1 and 2 reveal that the inherent trait of benevolent negotiators positively moderates the relationships between: (1) cooperative approach and time-saving negotiation outcome; and (2) dominating approach and deterioration of relationship negotiation outcome, respectively. Specifically, Model 1 (Model 2) can be described as follows: the degree of achieving a time-saving (deterioration of relationship) negotiation outcome based on the adoption of a cooperative (dominating) negotiating behaviour may be higher for negotiators with greater tolerance for being under-rewarded (i.e., benevolent negotiators). As mentioned, benevolent negotiators are the givers at the negotiating table. If benevolent negotiators adhere to norms of reciprocity such as engaging in and responding to cooperative behaviour in the course of negotiation, then the entire negotiation process will be smooth (Graham *et al.* 1994; Purdy *et al.* 2000; Rubin and Brown 1975) and the efficiency of the negotiation will be improved. Model 2 provides further evidence of the moderating effect of benevolent negotiators. They think more of giving than receiving (Rychlak 1973), and do not ignore the needs and expectations of their counterparts. Otherwise, a dysfunctional negotiation outcome results.

Equity sensitive negotiators

Equity sensitive negotiators prefer that their input/outcome ratio be equal to that of their counterpart(s) (Allen and White 2002; Sauley and Bedeian 2000). Model 3 reveals that the inherent trait of equity sensitive negotiators positively moderates the relationship between aggressive negotiating behaviour and deterioration of relationship negotiation outcome. That is, the degree of the deterioration of the relationship based on the adoption of aggressive negotiating behaviour may be higher for equity sensitive negotiators. According to Allen *et al.* (2004), equity sensitive negotiators insist on reaching a state of equity with regard to the outcomes they receive for the amount of inputs they expend compared to their counterparts doing similar work. If their input/output ratio is out of balance, then they will be motivated to get the ratio back into balance (Adams 1963, 1965; Allen *et al.* 2004). The findings of this study show that if these negotiators adopt an aggressive approach to restore their input/output ratio, then a climate of hostility and distrust develops.

Entitled negotiators

Entitled negotiators are takers who prefer to receive more than they give (Sauley and Bedeian 2000). They often compare their own input/outcome ratio with that of their counterpart(s). In the negotiation process, they may experience distress if they are unable to get a better deal than their negotiation counterparts (Huseman *et al.* 1987). This perception may act as a catalyst for the improvement of the quality of negotiation outcomes. The inherent trait of entitled negotiators is found to be a significant moderator of the relationships between the dominating and the cooperative approach and satisfactory negotiation outcomes (see Models 4, 5 and 6).

Practical implications and further research areas

The increasingly complex nature of construction projects has led to a substantial increase in the number of disputes and claims (Egan 1998; Latham 1994). In practice, negotiation is usually considered to be the most efficient dispute or claim resolution method in the construction industry. Using the data of experienced negotiators of construction disputes, this study applied the concept of equity sensitivity to the study of construction dispute negotiation to investigate how negotiators react in equitable and inequitable situations. This research has important implications, as previous research reveals that perceptions of equity affect negotiating behaviour (Mintu-Wimsatt 2005; Vecchio 1981). Most importantly, this study advances the present understanding of equity sensitivity and negotiation by extending its scope through investigating the relationship between negotiating behaviours and negotiation outcome. Specifically, it investigated the role of equity sensitivity in moderating this relationship. With this specific objective, an equity sensitivity construct was developed. This construct suggested that construction negotiators do not conform consistently to the norm of equity. It also appears to be relevant by explaining how negotiators actually perceive inputs and outcomes. These negotiators are most likely to be entitleds, who expect their input/outcome ratio to exceed that of their negotiating partner(s). This finding is in line with the proposition of Huseman *et al.* (1987) that 'negotiators [will] react consistently to specific, but different, preferences they have for the balance between their outcome/input ratio and that of a comparison other.' Second, using the equity sensitivity construct that was used to relate negotiating behaviour and negotiation outcomes, MMR revealed a total of six significant MMR models. These models explain how negotiators of construction disputes react to equitable or inequitable situations, and explain the moderating effects of these actions on behaviour–outcome relationships. Among the three major classes of negotiators, entitled negotiators were found to be versatile moderators who fostered satisfactory negotiation outcomes in construction dispute negotiation. The findings have important practical implications for managers in the construction industry. They need to be aware that the majority of negotiators in construction are entitleds, who will behave less cooperatively than the other classes of negotiators. Negotiators should also take this into account when they design strategies for their negotiations. The significant MMR models in this study can also help negotiators to choose an appropriate type of negotiating behaviour. For instance, as shown in Figure 9.3, Model 4 suggests that the adoption of the dominating approach would be appropriate for interactions with entitleds to achieve satisfactory negotiation outcomes. Model 3 suggests that the deterioration of a relationship may result from the use of an aggressive approach by equity sensitive negotiators. Taken together, the six significant MMR models suggest that the contingent adoption of negotiating behaviour or strategies is essential in construction dispute negotiation. Again, taking Model 4 as an example, this model reveals that the dominating approach is contingently effective for achieving satisfactory negotiation outcomes when used by entitled negotiators. In this connection, this study can also inspire negotiators in the construction industry to take a closer look at the effects that equity sensitivity appears to have on behavioural responses and outcomes. Finally, the findings of this study merit further research into equity theory in business negotiation. For example, future research could predict negotiators' options for bringing equity ratios into balance (Allen and White 2002; Greenberg 1990). This is known as equity restoration, and is critical to the creation of an equitable environment in construction dispute negotiation.

Summary

This chapter reports a study that examines the moderating effect of equity sensitivity on behaviour–outcome relationships in construction dispute negotiation. With the data obtained from business practitioners in the construction industry, an equity sensitivity construct was first developed, with a five-factor solution. This finding provided a more detailed classification of Entitleds, which indicated that negotiators of this type are probably less cooperative than the other types. They prefer to be over-rewarded and feel distress when they perceive that they are equitably rewarded or under-rewarded (Foote and Harmon 2006). One possible explanation for this finding is that the construction industry is often associated with the cutting of corners, defects, poor workmanship and disputes. In the event of disputes or claims, the different interests, needs or goals of groups/organisations are often incompatible, and negotiators endeavour to maximise their own benefits.

Acknowledgements

Special thanks to Miss Yee Man Law for collecting data for this study. Part of the content of this chapter has been published in Volume 137(5) of the

Journal of Construction Engineering and Management and is used with the permission from the American Society of Civil Engineers (ASCE).

References

Adams, J. S. (1963) Toward an understanding of inequity. *Journal of Abnormal and Social Psychology*, 67, 422–36.
Adams, J. S. (1965) Inequity in social exchange, in L. Berkowitz (ed.), *Advances in Experimental Psychology*, Vol. 2. San Diego, CA: Academic Press, 269–99.
Allen, R. S. and White, C. S. (2002) Equity sensitivity theory: a test of responses to two types of under-reward situations. *Journal of Managerial Issues*, 14(4), 435–51.
Allen, R. S., Biderman, M. and White, C. S. (2004) Emotional intelligence and its relation to equity sensitivity and responses to under-reward situations. *The Journal of Behavioral and Applied Management*, 5(2), 114–36.
Baron, R. A. (1990) Environmentally induced positive affect: its impact on self-efficacy, task performance, negotiation, and conflict. *Journal of Applied Social Psychology*, 20, 368–84.
Baron, R. A. and Richardson D. R. (1994) *Human Aggression*. New York: Plenum Press.
Bazerman, M. H., Curhan J. R., Moore, D. A. and Valley, K. L. (2000) Negotiation. *Annual Review of Psychology*, 51, 279.
Bennett, J. and Jayes S. (1995) *Trusting the Team: The Best Practice Guide to Partnering in Construction*. Reading: Centre for Strategic Studies in Construction, Reading Construction Forum.
Berry, W. D. and Feldman, S. (1985) *Multiple Regression in Practice*. Beverly Hills, CA: Sage.
Bing, M. N. and Burroughs, S. M. (2001) The predictive and interactive effects of equity sensitivity in teamwork-oriented organizations. *Journal of Organizational Behavior*, 22, 271–90.
Blakely, G. L., Andrews, M. C. and Moorman, R. H. (2005) The moderating effects of equity sensitivity on the relationship between organizational justice and organizational citizenship behaviors. *Journal of Business and Psychology*, 20(2), 259–73.
Cheung S. O. and Yeung Y. W. (1998) The effectiveness of the Dispute Resolution Advisor System: a critical appraisal. *The International Journal of Project Management*, 16(6), 367–74.
Cobb, L., Koppstein, P. and Chen, N. H. (1983) Estimation and moment recursion relations for multimodal distributions of the exponential family. *Journal of the American Statistical Association*, 78, 124–30.
Cohen, J., Cohen, P., West, S. G. and Aiken, L. S. (2003) *Applied Multiple Regression/Correlation Analysis for the Behavioral Science*. Mahwah, NJ: L. Erlbaum Associate.
Crane, T. G., Felder, J. P., Thompson, P. J., Thompson, M. G. and Sanders, S. R. (1999) Partnering Parameters. *Journal of Management in Engineering*, 15(2), 37–42.
Dabholkar, P., Johnson, W. and Cathey, A. (1994) The dynamics of long-term business-to-business relationship. *Journal of the Academy of Marketing Science*, 22, 130–45.

Dozzi, P., Hartman, F., Tidsbury, N. and Ashrafi, R. (1996) More-Stable Owner-Contractor Relationships. *Journal of Construction Engineering and Management*, 122(1), 30–35.

Egan, J. (1998) *Rethinking Construction*. London: Department of the Environment, Transport and the Regions.

Fang, D., Li, M., Fong, P. S. & Shen, L. (2004) Risks in Chinese construction market: contractors' perspective. *Journal of Construction Engineering and Management*, 130(6), 853–61.

Fava, J. L., and Velicer, W. F. (1992) The effects of overextraction on factor and component analysis. *Multivariate Behavioral Research*, 27(3), 387–415.

Fong, P. S. and Chu, L. (2006) Exploratory study of knowledge sharing in contracting companies: a sociotechnical perspective. *Journal of Construction Engineering and Management*, 132(9), 928–39.

Foote, D. A. and Harmon, S. (2006) Measuring equity sensitivity. *Journal of Managerial Psychology*, 21(2), 90–108.

Frazier, G.L. and Summers, J. (1984) Interfirm influence strategies and their application within distribution channels. *Journal of Marketing*, 48(Summer), 43–55.

Friedman, R.A., Currall, S.C. and Tsai, J.C. (2000) What goes around comes around: The impact of personal conflict style on work conflict and stress. *The International Journal of Conflict Management*, 11(1), 32–55.

Goering, E. (1997) Integration versus distribution in contract negotiations: an interaction analysis of strategy use. *The Journal of Business Communication*, 34, 383–400.

Goyal, M. (2004) An attitude based cooperative negotiation model, in *Proceedings of the Fourth International Conference on Hybrid Intelligent Systems*, 476–77.

Graham, J. L., Mintu, A. T. and Rodgers, W. (1994) Explorations of negotiation behaviors in ten foreign cultures using a model developed in the United States. *Management Society*, 40(1), 72–95.

Greenberg, J. (1990) Organizational justice: yesterday, today and tomorrow. *Journal of Management*, 16, 399–432.

Gulati, R. (1995) Does familiarity breed trust? The implications of repeated ties for contractual choice in alliances. *Academy of Management Journal*, 38(1), 85–112.

Hair, J. F., Anderson, R. E., Tatham, R. L. and Black, W. C. (1998) *Multivariate Data Analysis* (5th edn). Upper Saddle River, NJ: Prentice Hall.

Hartman, S. J. and Villere, M. F. (1990) A fair shake is not for everyone: Equity Theory revisited. *Leadership and Organization Development Journal*, 11(7), 1–4.

Hodgson, J. (1996) *Thinking on Your Feet in Negotiations*. London: Pitman Publishing.

Holt G. (1997) Construction research questionnaire and attitude measurement: relative index or mean. *Journal of Construction Procurement*, 3(2), 88–94.

Huseman, R. C., Hatfield, J. D. and Miles, E. W. (1985) Test for individual perceptions of job equity: some preliminary findings. *Perceptual and Motor Skills*, 61, 1055–64.

Huseman, R. C., Hatfield, J. D. and Miles, E. W. (1987) A new perspective on Equity Theory: the equity sensitivity construct. *The Academy of Management Review*, 12(2), 222–34.

Jaccard, J., Turrisi, R. and Choi, K. W. (1990) *Interaction Effects in Multiple Regression*. Thousand Oaks, CA: Sage.

John D. and Jack, O. (2005) *Cooperative Behavior and the Frequency of Social Interaction*. Pittsburgh: Department of Economics, University of Pittsburgh.

Kickul, J. and Lester, S. W. (2001) Broken promises:equity sensitivity as a moderator between psychological contract breach and employee attitudes and behaviour. *Journal of Business and Psychology*, 16(2), 191–16.

Kickul, J. Gundry, L. K. and Posig, M. (2005) Does trust matter? The relationship between equity sensitivity and perceived organizational justice. *Journal of Business Ethics*, 56, 205–18.

King, W. C. and Hinson, T. D. (1994) The influence of sex and equity sensitivity on relationship preferences, assessment of opponent, and outcomes in negotiation experiment. *Journal of Management*, 20, 605–24.

King, W. C. Jr., Miles, E. W. and Day, D. D. (1993) A test and refinement of the equity sensitivity construct. *Journal of Organizational Behavior*, 14(4), 301–17.

Latham, M. (1994) *Constructing the Team: Final Report by Sir Michael Latham*, Joint Review of Procurement and Contractual Arrangements in the United Kingdom Construction Industry. London: HMSO.

Leventhal, G. S. (1976) *Fairness in Social Relationships*. Morristown, NJ: General Learning Press.

Lewis-Beck, M. S. (1980) *Applied Regression: An Introduction*. Beverly Hills, CA: Sage.

Lim, R. G. and Carnevale, P. J. D. (1990) Contingencies in the mediation of disputes. *Journal of Personality and Social Psychology*, 58(2), 259–72.

Luck, D. J. and Rubin, R. S. (1987) *Marketing Research* (7th edn). Englewood Cliffs, N.J.: Prentice-Hall.

Luo, Y. (2002) Contract, cooperation, and performance in international joint ventures. *Strategic Management Journal*, 23, 903–19.

Maoz, I. (2005) Evaluating the communication between groups in dispute: equality in contact interventions between Jews and Arabs in Israel. *Negotiation Journal*, 21(1), 131–46.

McLoughlin, D. and Carr, S. C. (1997) Equity sensitivity and double demotivation. *The Journal of Social Psychology*, 137, 668–70.

Messick, D. M. and Sentis, K. (1983) Fairness, preference, and fairness biases, in D. M. Messick and K. S. Cook (eds), *Equity Theory: Psychological and Sociological Perspectives*. New York: Praeger, 166–89.

Miles, E., Hatfield, J. and Huseman, R. (1989) The equity sensitivity construct: potential implications for worker performance. *Journal of Management*, 15, 581–88.

Mintu-Wimsatt, A. (2005) Equity sensitivity and negotiation behaviors: a look at Mexican exporter. *Academy of Marketing Science Review*, 1, 1–11.

Mintu-Wimsatt, A. and Calantone, R. J. (1996) Exploring factors that affect negotiators' problem-solving orientation. *Journal of Business & Industrial Marketing*, 11(6), 61–73.

Monge, P., Fulk, J., Wilson, M., Gibbs, J. and Lee, B. (1997) The effects of cooperative & competitive physical environments and communication on negotiation outcomes in ultimatum and communication games. Available at: http://nosh.northwestern.edu/externalfunded/Steelcase97.pdf.

Mudrack, P. E., Mason, E. S. and Stepanski, K. M. (1999) Equity sensitivity and business ethics. *Journal of Occupational and Organizational Psychology*, 72, 539–60.

Mueller, L. S. and Clarke, L. D. (1998) Political-economic context and sensitivity to equity: differences between the United States and the transition economies of Central and Eastern Europe. *The Academy of Management Journal*, 41(3), 319–29.

Newcombe, R. (1996) Empowering the construction project team. *International Journal of Project Management*, 14(2), 75–80.

Nieuwmeijer, L. (1988) *Negotiation: Methodology and Training*. Pretoria: HSRC Press.

Nolan-Haley, J. M. (1992) *Alternative Dispute Resolution in a Nutshell*. St. Paul, Minnesota: West Publishing

Olekalns, M., Smith, P. L. and Kibby, R. (1996) Social value orientation and negotiation outcomes. *European Journal of Social Psychology*, 26, 299–313.

Plowman, K. D. (1998) Power in conflict for public relations. *Journal of Public Relations Research*, 10(4), 217–61.

Porter, M. E. (1979) How competitive forces shape strategy. *Harvard Business Review*, 57(2), 137–45.

Pruitt, D. (1981) *Bargaining Behavior*. New York: Academic Press.

Pruitt, D. G. and Kimmel, M. J. (1977) Twenty years of experimental gaming: critique, synthesis, and suggestions for future research. *Annual Review of Psychology*, 29, 363–92.

Purdy, J. M., Nye, P. and Balakrishnan, P. V. (2000) The impact of communication media on negotiation outcomes. *International Journal of Conflict Management*, 11(2), 162–87.

Putnam, L. L. and Jones, T. S. (1982) Reciprocity in negotiations: an analysis of bargaining interaction. *Communication Monographs*, 49(3), 171–91.

Rahim, M. A., Manger, N. R. and Shapiro, D. L. (2000) Do justice perceptions influence styles of handling conflict with supervisors? What justice perceptions, precisely? *The International Journal of Conflict Management*, 11(1), 9–31.

Rahim, M. A. (2002) Toward a theory of managing organizational conflict. *The International Journal of Conflict Management*, 13(3), 206–35.

Rinehart, L. M. and Page, T. J. (1992) The development and test of a model of transaction negotiation. *Journal of Marketing*, 56(4), 18–32.

Rubin, J. and Brown, B. (1975) *The Social Psychology of Bargaining and Negotiation*. New York: Academic Press.

Rychlak, J. F. (1973) *Introduction to Personality and Psychotherapy*. Boston: Houghton Mifflin.

Sauley, K. S. and Bedeian, A.G. (2000) Equity sensitivity: construction of a measure and examination of its psychometric properties. *Journal of Management*, 26(5), 885–910.

Schawarz, R. and Peutsch, C. (2001) Negotiation skills development, in S. Jevtic *et al. Proceedings of the Second Symposium: Extra Skills for Young Engineers ESYE 2001*. 15–18.

Sheppard, B. H., Blumenfeld-Jones, K. and Roth, J. (1989) Informal thirdpartyship: studies of everyday conflict intervention, in K. Kressel, D. G. Pruitt *et al.* (eds), *Mediation Research: The Process and Effectiveness of Third-Party Intervention*. San Francisco: Jossey Bass Publishers, 61–94.

Sherif, C. W., Sherif, M. and Nebergall, R. E. (1965) *Attitude and Attitude Change: The Social Judgment Involvement Approach*. Philadelphia: Saunders.

Shore, T., Sy, T. and Strauss, J. (2006) Leader responsiveness, equity sensitivity, and employee attitudes and behaviour. *Journal of Business and Psychology*, 21(2), 227–41.

Thompson, L. (1990) Negotiation behavior and outcomes: empirical evidence and theoretical issues. *Psychological Bulletin*, 108(3), 515–32.

Tornow, W. W. (1971) The development and application of an input-output moderator test on the perception and reduction of inequity. *Organizational Behavior and Human Performance*, 6, 614–38.

Turner, J. R. (2004) Farsighted project contract management: incomplete in its entirety. *Construction Management and Economics*, 22(1), 75–83.

Vecchio, R. (1981) A individual-differences interpretation of the conflicting predictions generated by equity theory and expectancy theory. *Journal of Applied Psychology*, 66, 470–81.

Walker, A. (2002) *Project Management in Construction* (2nd edn). Oxford: Blackwell Scientific.

Walton, R. and McKersie, R. (1965) *Behavioral Labor Theory of Labor Negotiations*. New York: McGraw-Hill Book Company.

10 Equity in construction contracting negotiation

II: A study of problem-solving approaches and satisfaction

Tak Wing Yiu and Chung Wai Keung

Problem-solving approaches

Contracting parties are encouraged to adopt problem-solving approaches (PSAs) to resolve disputes as they arise. PSAs involve negotiating behaviour and focus on solving mutual problems in a way that is accommodating, honest and unbiased. Negotiators who adopt such approaches are cooperative, integrative and information-exchange oriented (Graham 1986). The adoption of a PSA is regarded as the most effective way to resolve organisational conflict (Calantone *et al.* 1998; Ghauri 1986; Nickerson and Zenger 2004; Rahim 2002) and build long-term relationships (Ganesan 1994). Such approaches provide a strong impetus to understand cultural differences (Dabholkar *et al.* 1994; Eliashberg *et al.* 1995) and simulate actual negotiation behaviour (Graham 1985; Graham and Andrews 1987). PSAs have been widely investigated in the field of business negotiation (e.g., Adler and Graham 1989; Graham 1985, 1986; Mintu-Wimsatt 2005). These studies generally agree that negotiators should adopt a PSA for their first move. The first move sets the tone for the entire negotiating process. In addition, as bargaining interaction is often characterised by reciprocal exchanges (Alexander *et al.* 1991; Goering 1997; Putman and Jones 1982), integrative (or distributive) messages tend to be matched by integrative (or distributive) responses (Mintu-Wimsatt 2005). Studies show that when one party adopts a PSA, his or her negotiating partner is likely to adopt one as well (Calantone *et al.* 1998; Graham 1986). Negotiators in the construction industry often face tough negotiation processes that involve the diverse interests of the various contracting parties, including clients, contractors and sub-contractors. The adversarial and non-cooperative culture of this industry may raise particular challenges for negotiators who would like to adopt PSAs. This study addresses the way in which PSAs can be facilitated. We first conceptualise the negotiator's sensitivity to perceived equity or inequity (hereafter, equity sensitivity) by applying equity sensitivity theory at the individual level. Three classes of negotiators (hereafter, equity sensitivity groups) are identified according to their different responses to perceived equity or inequity. Next, the differences in the perception of the adoption of PSAs and level of negotiation satisfaction among the three equity sensitivity groups are investigated. As negotiation in the construction industry often involves the resolution of disputes or claims, a better understanding of those differences will be beneficial to negotiators who would like to adopt PSAs. The findings of this study will also help the corporate managers of construction organisations to recognise and understand the importance of the equity sensitivity of their negotiators.

The hypotheses

Equity sensitivity theory has been introduced in Chapter 9. Based on the inherent traits of three classes of negotiators – benevolents, equity sensitives and entitleds – we hypothesise that their perceptions of the adoption of PSAs and levels of negotiation satisfaction are different. Benevolent negotiators are predicted to take the initiative during the course of negotiations to work together with their negotiating partners to share information about needs and preferences. If their counterparts reciprocate this behaviour, then they will obtain satisfaction. This supposition is supported by the findings of Calantone *et al.* (1998) and Campbell *et al.* (1988) that negotiators often express a high level of satisfaction when a PSA is used by both parties at the negotiating table. Calantone *et al.* (1998) and Graham (1986) identify four types of satisfaction: (1) satisfaction with negotiation outcomes; (2) satisfaction relative to pre-negotiation expectations; (3) satisfaction with the level of organisational profit; and (4) satisfaction with negotiating performance. These provide the basis for evaluating the success or failure of negotiations and are the factors that lead to desired negotiation outcomes (Calantone *et al.* 1998).

Based on the foregoing discussion, we put forward the following hypotheses.

H1: *If negotiators belong to different equity sensitivity groups, then their perceptions of adopting a PSA to solve mutual problems will differ.*

H2: *If negotiators belong to different equity sensitivity groups, then their perceptions of the adoption of an accommodating PSA will differ.*

H3: *If negotiators belong to different equity sensitivity groups, then their perceptions of the adoption of an honest PSA will differ.*

H4: *If negotiators belong to different equity sensitivity groups, then their perceptions of the adoption of an unbiased PSA will differ.*

H5: *If negotiators belong to different equity sensitivity groups, then their levels of satisfaction with negotiation outcomes will differ.*

H6: *If negotiators belong to different equity sensitivity groups, then their levels of satisfaction with negotiation outcomes relative to pre-negotiation expectations will differ.*

H7: *If negotiators belong to different equity sensitivity groups, then their levels of satisfaction with the level of organisational profit will differ.*

H8: *If negotiators belong to different equity sensitivity groups, then their levels of satisfaction with negotiating performance will differ.*

Hypotheses testing

Data collection

The primary objective of this study was to investigate how the perception of the adoption of PSAs and the level of negotiation satisfaction varies among benevolent, equity-sensitive and entitled negotiators. To achieve this objective, a questionnaire was designed to collect data from target respondents in Hong Kong. These were professionals in construction organisations, namely, project managers, surveyors, architects and engineers, whose responsibilities include negotiating claims and disputes for their organisations. The target respondents were selected from construction companies that were registered in the Builder Directory and from the Web pages of professional institutes. Those who agreed to participate in the study were sent a questionnaire survey by post, fax or email, according to their choice. The questionnaire incorporated the following measures.

Measures of equity sensitivity

The Equity Sensitivity Instrument (ESI) is a well-established measure of equity sensitivity that was developed by Huseman *et al.* (1985, 1987). ESI is used to identify differences in the way individuals view inequitable situations (Foote and Harmon 2006). This five-item forced-distribution scale gauges the preferences of respondents for inputs versus outcomes during negotiation. Each item comprises two statements: an entitled and a benevolent response (Sauley and Bedeian 2000). Respondents are asked to indicate their preference by apportioning 10 points between the two statements. Foote and Harmon (2006) pointed out four attributes of the ESI: (1) it demonstrates a high degree of internal reliability; (2) it appears to be unidimensional; (3) all of its items that are related to the input/outcome exchange are derived from equity theory; and (4) it may be more representative of the general public as it was developed using a non-student sample. The ESI has been widely used in previous equity sensitivity studies (Allen and White 2002; Kickul and Lester 2001; King *et al.* 1993; Miles *et al.* 1989; Miles *et al.* 1994; Mintu-Wimsatt 2003; O'Neill and Mone 1998; Patrick and Jackson 1991; Wheeler 2002). For these reasons, we chose the ESI to measure equity sensitivity with modifications to suit the construction context. The ESI results are covered in the Findings and Discussion section.

PSAs and negotiation satisfaction

PSAs and negotiation satisfaction were measured using the problem-solving approach dimensions (hereafter, PSA dimensions) developed by Graham (1986). PSA dimensions have been used extensively to investigate negotiation behaviour (Adler and Graham 1989; Graham *et al.* 1994; Mintu-Wimsatt 2005; Mintu-Wimsatt and Graham 2004). They comprise two sets of questions. In the first set, the perception of respondents of *solving mutual problems* is measured by the question: 'Do you feel that you are more interested in solving mutual problems or are you more self-interested?' Respondents are asked to rate the degree of their self-interest/interest in solving mutual problems on a Likert scale that ranges from 1 'completely self-interested' to 7 'completely interested in solving mutual problems'. Three other major PSAs, *accommodating, honest* and *unbiased* approaches, are rated using 7-point, self-reported, itemised-category scales with opposing adjectives as anchors (Graham *et al.* 1994). These scales range from 1 'exploitative' to 7 'accommodating', from 1 'deceptive' to 7 'honest' and from 1 'biased' to 7 'unbiased', respectively. Negotiation satisfaction is measured by four questions: (a) 'How satisfied were you with the negotiation outcome(s)?' (b) 'How satisfied were you with the negotiation outcome(s) relative to your pre-negotiation expectations?' (c) "How satisfied were you with your organisational profit level?' and (d) 'How satisfied were you with your performance during the negotiation process?' These questions are evaluated using a 7-point Likert scale that ranges from 1 'low degree of satisfaction' to 7 'high degree of satisfaction'. The content of the original PSA dimensions was slightly modified to suit the construction context. For example, as construction negotiators represent the interests of profit-making organisations (Loosemore 1999), the question 'How satisfied were you with your individual profit level' was modified to read 'How satisfied were you with your organisational profit level?'.

To test the hypotheses, one-way analysis of variance (ANOVA) was performed to determine whether there were any significant differences in perception of PSA adoption and level of negotiation satisfaction among the three classes of negotiators. ANOVA is a type of statistical analysis that is used to evaluate the equality of means, such as mean differences, of a single intervally-scaled outcome across two or more groups (Thompson 2006). The significant ANOVA results were then followed up using Duncan's Multiple Range Test (hereafter, Duncan's test), a widely used procedure for determining the source of any significant differences (King *et al.* 1993; Miles *et al.* 1994; Montgomery 1996).

Findings and discussion

A total of 165 questionnaires were distributed to the target respondents, 90 of whom returned completed questionnaires, giving a response rate of

Table 10.1 Composition of respondents (by organisation type)

Type of organisation	Sent (nos.)	Returned (nos.)	Response rate (%)
a. Private developer	20	8	40.0
b. Government agency	20	15	75.0
c. Consultancy firm	80	40	50.0
d. Contractor	45	27	60.0
Total	165	90	54.5

54.50 per cent. The composition by organisation type of the respondents, 72 per cent of whom had more than five years of experience in construction negotiation, is presented in Table 10.1.

ESI scores

The ESI scores are the sum of the points allotted to the two benevolence statements (King *et al.* 1993), and range from 6 to 47 (mean = 23.83, SD = 7.28) out of a possible range of 0 to 50. To divide the sample into the three equity sensitivity groups, the sample-specific breakpoints used by Huseman *et al.* (1985), Miles *et al.* (1989, 1994) and King *et al.* (1993) were adopted, as the unique characteristics of any particular sample may influence ESI responses (King *et al.* 1993). These breakpoints were approximately ± half a standard deviation from the mean of the entire sample (King *et al.* 1993), and the resulting three categories are presented in Table 10.2. Respondents with a score of 20 or less were classified as entitleds (n = 29, mean = 16.03, SD = 3.59); those with scores between 21 and 27 as equity sensitives (n = 35, mean = 23.94, SD = 2.10); and those with scores of 28 or greater as benevolents (n = 26, mean = 32.38, SD = 4.63). The distribution of these three classes of negotiators by organisation, including private developer, government agency, consultancy firm or contractor, is shown in Table 10.3.

Half of the negotiators in construction organisations (45 out of the 90 respondents) were found to be entitleds, those known as takers at the negotiating table, which may perhaps be a root cause of the adversarial and

Table 10.2 Breakpoints of the ESI scores (Huseman *et al.* 1985; Miles *et al.* 1989)

Three classes of negotiators	Breakpoints of ESI scores
Entitleds	0–20
Equity sensitives	21–27
Benevolents	28–50

Table 10.3 The three classes of negotiators (by organisation type)

| | Three classes of negotiators | | |
	Entitleds	Equity sensitives	Benevolents
Private developer	5	3	0
Government agency	8	6	1
Consultancy firm	18	13	9
Contractor	14	3	10
Total	45	25	20

non-cooperative nature of the construction industry. The majority of the respondents work for consultancy firms and thus contractually represent the client. The general aim of negotiators of such organisations is to maximise client's profit in construction projects, and therefore their first priority is the client's interests. Entitled negotiators will try to employ commercial strategies to fight for the interests of their clients at the negotiating table. Interestingly, a number of benevolents were found among the negotiators employed by contractors, which may reflect the high turnover rate of construction projects. These negotiators might have little motivation to advance the interests of their clients but rather prefer to reach an agreement with their negotiating partners as quickly as possible so as not to lose their baseline.

The three equity sensitivity groups

Having separated the sample into the three classes of negotiators, a series of one-way ANOVA were conducted to determine how the equity sensitivity groups differed in their perception of the adoption of PSAs. The significant ANOVA results were again followed by the use of Duncan's test to isolate the source of the differences (King *et al.* 1993). Tables 10.4 and 10.5 show the ANOVA and the Duncan's test results respectively.

H1, H2, H3 and H4 predicted differences among the three types of negotiators in their perception of the adoption of a PSA to solve mutual problems ($F = 17.273$, $p < 0.01$) and the accommodating ($F = 8.534$, $p < 0.05$), honest ($F = 17.974$, $p < 0.01$) and unbiased PSAs, respectively ($F = 5.790$, $p < 0.01$). The ANOVA results are strongly significant, and the Duncan's test results identify two significantly different groups (Table 10.5). The perception of the adoption of a PSA to solve mutual problems differs significantly among the three types of negotiators, thus providing support for H1. The mean benevolent score, (5.27), is 26 per cent and 64 per cent greater than the equity sensitive and entitled scores of 4.17 and 3.21, respectively. This result is unsurprising, given the inherent traits of benevolent negotiators. This serves also as statistical support for the proposition of

Table 10.4 Results of one-way ANOVA

		SS*	df	F	Sig.
Solving mutual problems	Between groups	58.31	2	17.273	.000**
	Within groups	146.85	87		
	Total	205.16	89		
Accommodating	Between groups	22.85	2	8.534	.000**
	Within groups	116.44	87		
	Total	139.29	89		
Honest	Between groups	48.24	2	17.974	.000**
	Within groups	116.75	87		
	Total	164.99	89		
Unbiased	Between groups	19.31	2	5.790	.004**
	Within groups	145.09	87		
	Total	164.40	89		

(left axis: *Type of PSA*)

Notes: *p < 0.05; **p < 0.01.
Cronbach's alpha of PSA dimensions: 0.768.

Table 10.5 Results of Duncan's Multiple Range Test

	Cell means			
	ENT (n = 29)	EQS (n = 35)	BEN (n = 26)	Duncan's Test*
Solving mutual problems	3.21	4.17	5.27	ENT EQS BEN
Accommodating	3.76	4.23	5.04	ENT EQS BEN
Honest	4.34	4.69	6.12	ENT EQS BEN
Unbiased	3.93	4.43	5.12	ENT EQS BEN

Notes: BEN: Benevolent negotiator; EQS: Equity-sensitive negotiator; ENT: Entitled negotiator.
*Groups connected by a solid line are not significantly different from one another.

Mintu-Wimsatt (2005) and King *et al.* (1993) that this type of negotiator has a propensity to solve mutual problems during the negotiation process. This result is also empirical evidence that this proposition is applicable to the construction industry notwithstanding the confrontational culture (CIRC 2001; Egan 1998). Benevolent negotiators are more likely to create an environment that fosters cooperation and teamwork, which can obviously lead to the achievement of mutually beneficial outcomes. No significant differences were found between entitled and equity-sensitive negotiators in their perceptions of the adoption of an accommodating PSA, but benevolent negotiators were significantly different from both, which provides partial support for H2. According to the dual-concern model (Thomas 1976), the accommodating approach is characterised by a low degree of concern for oneself and a high degree of concern for others, which, to a certain extent, is consistent with the behavioural traits of the benevolent negotiator, who is also known as the giver. It is thus expected that these negotiators will adopt an accommodating PSA more frequently than either their equity-sensitive or entitled counterparts.

The results of Duncan's test also reveal that benevolent negotiators are significantly different from the other two types in their perception of the adoption of honest and unbiased PSAs. The cell means presented in Table 10.5 show that benevolent negotiators have a significantly greater preference for these two types of strategies. Table 10.5 also shows that the scores obtained for the extent of adoption of an honest PSA remain above the median of the 7-point scale across the three classes of negotiators, which indicates the general practice of honesty in negotiation in the construction industry. These negotiations are regulated by contracts that define the obligations and rights of the contracting parties (Cheung *et al.* 2008), and deceptive strategies, such as hiding information, shaping impressions or making false statements (Provis 2000), are seen as impeding *effective* negotiation. Most importantly, a dishonest approach could lead to the tarnishing of reputations and ongoing business relationships, which would restrict future tender opportunities.

With regard to the level of negotiation satisfaction, H5 and H6 are partially supported. The ANOVA results presented in Table 10.6 show significant differences in the level of satisfaction with negotiation outcomes ($F = 4.006$, $p < 0.05$) and satisfaction with negotiation outcomes relative to pre-negotiation expectations ($F = 5.049$, $p < 0.05$) among the three classes of negotiators ($F = 4.006$, $p < 0.05$). The results of Duncan's test show that entitled negotiators are significantly different from their equity-sensitive and benevolent counterparts, but no significant differences are found between the latter two. Benevolent and equity-sensitive negotiators thus

Table 10.6 Results of one-way ANOVA

		SS	df	F	Sig.
Satisfaction with negotiation outcomes	Between groups	6.011	2	4.006	.022*
	Within groups	65.278	87		
	Total	71.289	89		
Satisfaction relative to pre-negotiation expectations	Between groups	12.209	2	5.049	.008**
	Within groups	105.180	87		
	Total	117.389	89		
Satisfaction with organisational profit level	Between groups	7.273	2	2.569	.082
	Within groups	123.182	87		
	Total	130.456	89		
Satisfaction with negotiating performance	Between groups	1.065	2	.522	.595
	Within groups	88.724	87		
	Total	89.789	89		

Note: * $p < 0.05$; ** $p < 0.01$

have a significantly higher level of negotiation satisfaction than entitled negotiators. This finding is consistent with those of Huseman *et al.* (1985), King *et al.* (1993) and Graham *et al.* (1994). According to these studies, higher level of satisfaction among benevolents and equity sensitives is the result of the adoption of PSAs in the course of negotiation. Finally, H7 and H8 are not supported, as no significant differences are found among the three classes of negotiators in terms of the level of their satisfaction with their organisational profit level and performance at the negotiating table. Table 10.7 shows that the cell means for these two items remain above the median on the 7-point scale, which implies that all three types of negotiators have achieved satisfaction in these two respects.

Application of the results

This finding is particularly relevant to management of construction contracting organisations because it suggests that their negotiators can be differentiated by their perception of the adoption of a PSA. Creating the right match between a negotiator's inherent equity sensitivity traits and the negotiation agenda would enhance the effectiveness of the negotiation process. This study provides insight to elaborate this proposition, and suggests the following guidelines for corporate managers of construction organisations in choosing negotiators.

1   This study revealed that benevolent negotiators are more likely to want to solve mutual problems. They are instrumental in creating a cooperative environment and fostering team spirit. Corporate managers should make use of these strengths of the benevolent negotiator to sustain long-term business relationships with their partners, which is particularly

Table 10.7 Duncan's Multiple Range Test results on negotiation satisfaction

	Cell means			
	ENT (n = 29)	EQS (n = 35)	BEN (n = 26)	Duncan's Test*
Satisfaction with negotiation outcomes	4.72	5.20	5.35	ENT EQS BEN
Satisfaction relative to pre-negotiation expectations	4.14	4.66	5.08	ENT EQS BEN
Satisfaction with organisational profit level	4.28	4.31	4.92	ENT EQS BEN
Satisfaction with negotiating performance	4.97	4.47	4.77	ENT EQS BEN

Notes: BEN: Benevolent negotiator; EQS: Equity-sensitive negotiator; ENT: Entitled negotiator.
*Groups connected by a solid line are not significantly different from one another.

important in the current harsh economic environment. Sustaining a competitive advantage and preserving good client–contractor relationships are crucial to construction organisations.

2   This study showed that entitled negotiators have relatively lower levels of negotiation satisfaction compared to their benevolent and equity-sensitive counterparts. Corporate managers should be aware of the implication of sending entitled negotiators to the negotiating table if their aim is to repair a relationship with negotiating partners and solicit their cooperation.

3   Corporate managers should also be aware of the questionable effectiveness of assigning benevolent negotiators to promote corporate interests. This type of negotiator is regarded as a giver at the negotiating table, and tends to adopt an accommodating PSA more frequently than do the other types.

Summary

The adoption of a problem-solving approach (PSA) is one of the most effective ways of mitigating the typically adversarial and non-cooperative nature of negotiations in the construction industry. Previous research strongly suggests that the adoption of a PSA by one negotiating party will trigger reciprocal behaviour among the negotiating partners, prompting them to use a similar approach. In construction, however, negotiators are often faced with a tough negotiation process that involves diverse interests among the contracting parties. This situation may pose particular difficulties for negotiators who would like to adopt an equitable PSA. To find a way to overcome these difficulties, this study applies equity sensitivity theory to examine how a negotiator's sensitivity to perceived equity or inequity varies with his or her perception of the adoption of a PSA in construction negotiations. Three types of negotiators, benevolent (known as 'givers'), equity sensitive and entitled (known as 'takers'), are identified among a sample of negotiators, with the results suggesting that most negotiators are the entitled type. Benevolent negotiators are found to have a significantly greater preference for the adoption of PSAs compared to the other types of negotiators. The findings of this study are particularly relevant to the corporate managers of construction organisations, as they suggest that the right match between a negotiator's inherent equity sensitivity traits and the negotiation agenda can enhance the effectiveness of the negotiating process.

Acknowledgements

Special thanks to Miss Kit Ling Wong for collecting data for this study. Part of the content of this chapter has been published in Volume 27(1) of the *Journal of Management in Engineering* and is used with the permission from the American Society of Civil Engineers (ASCE).

References

Adler, N. J. and Graham, John L. (1989) Cross-cultural interaction: the international comparison fallacy. *Journal of International Business Studies*, 515–37.

Alexander, J. F., Schul, P. L. and Bakaus, E. (1991) Analyzing interpersonal communications in industrial marketing negotiations. *Journal of the Academy of Marketing Science*, 19, 129–39.

Allen, R. S. and White, C. S. (2002) Equity sensitivity theory: a test of responses to two types of under-reward situations. *Journal of Managerial Issues*, 14(4), 435–51.

Calantone, R. J., Graham, J. L. and Mintu-Wimsatt, A. (1998) Problem-solving approach in an international context: antecedents and outcome. *International Journal of Research in Marketing*, 19–35.

Campbell, N., Graham, J., Jolibert, A. and Meissner, H. G. (1988) A comparison of marketing negotiations in France, Germany, the United Kingdom, and the United States. *Journal of Marketing*, 52, 46–62.

Cheung, S. O., Wong, W. K., Yiu, T. W. and Kwok, T. W. (2008) Exploring the influence of contract governance on construction dispute negotiation. *Journal of Professional Issues in Engineering Education and Practice*, 134(4), 391–98.

Construction Industry Review Committee (CIRC) (2001) *Construct for Excellence*. Hong Kong Special Administrative Government.

Dabholkar, P., Johnston, W. and Cathey, A. (1994) The dynamics of long-term business-to-business exchange relationships. *Journal of the Academy of Marketing Science*, 22, 130–45.

Egan, J. (1998) *Rethinking Construction*. London: Department of the Environment, Transport and the Regions.

Eliashberg, J., Lilien, G. and Kim, N. (1995) Searching for generalizations in business marketing negotiation. *Marketing Science*, 14, G47–G60.

Foote, D. A. and Harmon, S. (2006) Measuring equity sensitivity. *Journal of Managerial Psychology*, 21(2), 90–108.

Ganesan, S. (1994) Determinants of long-term orientation in buyer-seller relationships. *Journal of Marketing*, 58, 1–19.

Ghauri, P. (1986) Guidelines for international marketing negotiations. *International Marketing Review*, 72–81.

Goering, E. (1997) Integration versus distribution in contract negotiations: an interaction analysis of strategy use. *Journal of Business Communication*, 34, 383–400.

Graham, J. (1985) Cross-cultural marketing negotiations: a laboratory experiment. *Marketing Science*, 4, 130–46.

Graham, J. (1986) The problem-solving approach to negotiations in industrial marketing. *Journal of Business Research*, 14, 549–66.

Graham, J. L. and Andrews, D. (1987) A holistic analysis of cross-cultural business negotiations. *Journal of Business Communication*, 24(4), 63–77.

Graham, J., Mintu, A. and Rodgers, W. (1994) Explorations of negotiation behaviors in ten foreign cultures using a model developed in the United States. *Management Science*, 40, 72–95.

Huseman, R. C., Hatfield, J. D. and Miles, E. W. (1985) Test for individual perceptions of job equity: some preliminary findings. *Perceptual and Motor Skills*, 61, 1055–64.

Huseman, R., Hatfield, J. and Miles, E. (1987) A new perspective on equity theory: the equity sensitivity construct. *Academy of Management Review*, 12, 222–34.

Kickul, J. and Lester, S. W. (2001) Broken promises: equity sensitivity as a moderator between psychological contract breach and employee attitudes and behavior. *Journal of Business and Psychology*, 16, 191–217.

King, W., Miles, E. and Day, D. (1993) A test and refinement of the equity sensitivity construct. *Journal of Organizational Behavior*, 67, 133–42.

Loosemore, M. (1999) Bargaining tactics in construction disputes. *Construction Management and Economics*, 17, 177–88.

Miles, E., Hatfield, J. and Huseman, R. (1989) The equity sensitivity construct: potential implications for worker performance. *Journal of Management*, 15, 581–88.

Miles, E., Hatfield, J. and Huseman, R. (1994) Equity sensitivity and outcome importance. *Journal of Organizational Behavior*, 15, 585–96.

Mintu-Wimsatt, A. (2003) King and Miles' equity sensitivity instrument: a cross-cultural validation. *Psychological Reports*, 92, 23–26.

Mintu-Wimsatt, A. (2005) Equity sensitivity and negotiation behaviors: a look at Mexican exporters. *Academy of Marketing Science Review*. Available: www.amsreview.org/articles/wimsatt01-2005.pdf.

Mintu-Wimsatt, A. and Graham, J. (2004) Testing a negotiation model on Canadian anglophone and Mexican exporters. *Journal of the Academy of Marketing Science*, 32(3), 345–56.

Montgomery, D. (1996) *Design and Analysis of Experiments* (4th edn). New York: Wiley.

Nickerson, J. A. and Zenger, T. R. (2004) A knowledge-based theory of the firm: the problem-solving perspective. *Organization Science*, 15, 617–32.

O'Neil, B. S. and Mone, M. A. (1998) Investigating equity sensitivity as a moderator of relations between self-efficacy and workplace attitudes. *Journal of Applied Psychology*, 83(5), 805–16.

Patrick, S. and Jackson, J. (1991) Further examination of the equity sensitivity construct. *Perceptual Motor Skills*, 73, 1091–106.

Provis, C. (2000) Honesty in negotiation. *Business Ethics: A European Review*, 9(1), 3–12.

Putnam, L. L. and Jones, T. S. (1982) Reciprocity in negotiations: an analysis of bargaining interaction. *Communication Monographs*, 49, 171–91.

Rahim, M. A. (2002) Toward a theory of managing organizational conflict. *International Journal of Conflict Management*, 13(3), 206–35.

Sauley, K. S. and Bedeian, A. G. (2000) Equity sensitivity: construction of a measure and examination of its psychometric properties. *Journal of Management*, 26(5), 885–910.

Thomas, K. W. (1976) Conflict and conflict management, in M. D. Dunnette (ed.), *Handbook of Industrial and Organizational Psychology*. Chicago, IL: Rand McNally, 889–935.

Thompson, B. (2006) *Foundations of Behavioral Statistics*. New York: Guilford.

Wheeler, K. G. (2002) Cultural values in relation to equity sensitivity within and across cultures. *Journal of Managerial Psychology*, September/October, 612–27.

Index

Adam's equity theory (ET) 163–4; benevolent negotiators 163–4, 165, 168, 175, 184, 188–90, 190–1, 191–2; entitled negotiators 164, 167, 168, 176, 187–8, 190–1, 191–2; equity sensitive negotiators 164, 168, 190–1; individual differences 163; individuals' response to under-reward situations 163

Adenfelt, M. 22

affect-based trust 40, 41, 52, 53; being thoughtful 44, 47; emotional investments 44, 47; path coefficients results 51

aggressive approach to negotiation 169, 170

Aguinis, H.: description of MMR 108

Akaike Information Criterion (AIC) 133, 134

Al-Ghassani, A.: knowledge management through PMS 114

Allen, R. S.: equity sensitivity theory 162

Analysis of Moment Structures 5.0 (AMOS) 29, 88

Ankrah, N. A. 23

anticipation of project performance (PP) 126–7; comparison to pre-determined project goals 127; splitting factors 137

Anumba, C.: knowledge management through PMS 114

Aouad, G.: project monitoring systems (PMS) 125

Argyris, J. L.: conceptual model 86; definition of OL 79; types of OL 83

artefacts 21–2, 23

Ataoglu, T.: contractors' feedback and learning 126

attention to project monitoring feedback (AF) 125–6; allocation of resources 125; Balanced Scorecard (BSC) 125; bimodality 128; communication of contractors 125; dependent on contractor's aspiration of performance 137; KPI 125; pre-determined goals for contractors 125; purpose of 125

Balanced Scorecard (BSC) 26–7, 120; attention to project monitoring feedback (AF) 125; construction safety performance 125

Barrett, P.: innovation in the construction industry 34

Bayes Information Criterion (BIC) 133, 134

Bedeian, A. G.: equity sensitivity 165

behaviour: behavioural flexibility 146; catastrophic change in contracting behaviour 146–7; data collection methods for study 148–50; flexibility 148; satisficing 127; sudden change in 123–4; *see also* construction conflict behaviour

behavioural flexibility 146, 148, 149, 155; relationship with construction conflict level and tension level 148–57

behaviour learning 83

benchmarking: in Hong Kong 50

benevolent negotiators 163–4, 165, 168, 175, 184; PSAs and 188–9, 190–1, 191–2

Bennis, W.: definition of OL 79

bids: competitiveness of bidding 102; evaluation 50

bimodality 128; construction conflict behaviour 147

Milton Keynes UK
Ingram Content Group UK Ltd.
UKHW040101071024
449327UK00019B/710